Precision Farming and Protected Cultivation

NIPA® GENX ELECTRONIC RESOURCES & SOLUTIONS P. LTD.
New Delhi-110 034

www.nipaers.com
Browse, Search, Read & Buy...
Print Books
eChapters
eBooks
Publishers
Forthcoming
Subject Catalogues
New Titles
Online Resources on:
✓ Current Affairs
✓ Reasoning, Logic & Aptitude
✓ Competitive Examinations
✓ English Language Lab
✓ Writing & Pronunciation Tools
✓ Personality Development
✓ Online Programmes for Professionals
✓ Interview Preparation
Pay Using
PhonePe
Google Pay
paytm
PayPal
UPI
CC Avenue

Precision Farming and Protected Cultivation

Prof. Brahma Singh
Padma Shri Awardee
Former President Indian Society for Protected Cultivation
Former Director, Life Sciences, DRDO, New Delhi

NIPA® GENX ELECTRONIC RESOURCES & SOLUTIONS P. LTD.
New Delhi-110 034

nipa®

NIPA® GENX ELECTRONIC
RESOURCES & SOLUTIONS P. LTD.

101,103, Vikas Surya Plaza, CU Block
L.S.C. Market, Pitam Pura, New Delhi-110 034
Ph : +91 11 27341616, 27341717, 27341718
E-mail: newindiapublishingagency@gmail.com
web: www.nipabooks.com

For customer assistance, please contact
Phone: + 91-11-27 34 17 17
Fax: + 91-11- 27 34 16 16
E-Mail: feedbacks@nipabooks.com

ISBN: 978-81-19254-38-5

This is the Paperback edition of Harback edition

NIPA® also publishes books in a variety of electronic formats. Some content that appears in print may not be available in electronic books, and vice versa.

Composed and Designed by NIPA®.

Preface

Farming or agriculture is being done progressively since human started colonizing and preferred settlement rather shifting from place to place. Over the large periods agriculture or food, feed, fodder and fuel production is improving decades after decades if not century after century. Agro-technology of a preceding decade is different than the succeeding and present decade imbibing new and new technologies confined to tilling the land on large scale using common sense, science and rich experience gained in the past. Like this agriculture world over progressed up to the eighteenth century when mechanization of agriculture became popular. This mechanization headed towards precision, alongwith precision in every crop production activity. That is how the precision agriculture started using advanced state-of-the art technologies for right agriculture operation at right time and in right quantity to get maximum quality produce. Precision farming is infancy in India except laser levelling and drip irrigation or fertigation. Almost at the same time interest in environment control or modification of crops started to have early produce, more produce, off season produce on a limited piece of land. This was done by using improvised plant production structures, devices or techniques to have better return in comparison to conventional agriculture. The onset of climate change has necessitated promotion of this type of agriculture, popularly known as protected cultivation or greenhouse cultivation. The interest in greenhouse cultivation or development of greenhouse technology multiplied with the passage of time and in 20th century followed by 21st century. Protected cultivation or greenhouse technology or technologies have become very attractive and lot many research and innovations have given birth to hi-tech greenhouse technology and low –tech greenhouse technology. Hi-tech greenhouse technology mainly is used for horticulture crops, where most of the parameters of environment and inputs are controlled though automation and mechanization with sensors and artificial intelligence etc to create most suitable environment for the plant to express its potential in terms of quality and quantity. Off late instead of horizontal production vertical production or farming is becoming popular in certain crops particularly in big cities helping them in local food security or urban farming. The endless engineering inputs in precision and greenhouse technologies for production of large number of horticulture crops are visible and requires close collaboration among horticulturists and several specialized engineers to take these technologies for further heights.

In this book an attempt has been made to cover the course contents prescribed for the course "Precision Farming and Protected Cultivation" for undergraduate students by the competent authority. Complicated and difficult to understand technologies, terminology and gadgets have been explained in a simple and easily understandable language in the hope that the book would be useful not only to students but teachers and farmers. The available literature on the subject has been scanned, simplified and presented keeping in view the undergraduate students as users.

In writing of this book I have been assisted by my several colleagues particularly Dr Pritam Chandra, Former Director, ICAR-CIAE, Bhopal, M.P., Dr T. Jankiram, ADG (Hort.), ICAR headquarter, New Delhi, Dr Sanjay Kumar Dwivedi, Director, Defence Research Labortory, DRDO, Tezpur, Assam and Dr T. Stobdon, Deputy Director, DRDO-DIHAR, Leh (J&K), Dr TBS Rajput and Dr Neelam Patel, WTC, IARI, New Delhi, Dr Naved Sabir, Dr M. Hasan and Dr A.K. Singh, Principal Scientists, CPCT, IARI, New Delhi, Dr. P.L. Saroj, Director, Central Instititue of Arid Horticulture, Bikaner, Rajasthan, Dr D.K. Singh, Prof. Vegetable Science, College of Agriculture, GBPUAT, Pant Nagar, Udham Singh Nagar and Dr Ashish Yadav, Senior Scientist, ICAR- National Organic Farming Research Institute, Tadong, Gangtok, Sikkim. I express sincere thanks for their unconditional and timely support.

I am indebted for co-operation by my wife Smt Saroj Singh and other family members Ms Shailaja Saini, Ms Bobby Singh, Ms Neelu Singh, Ms Rachna Singh, Dr Pritam Singh, Mr. Gaurav Bansal, Mr. Ishaan Singh and Mr. Aditya Singh who never disturbed me during the period of writing this book except monitoring the progress with encouraging words.

31-05 -2019
New Delhi

Brahma Singh
Padma Shri Awardee

Contents

Part I: Precision Farming

1

Precision Farming

Precision farming is also known as precision agriculture. A farmer is not only expected to produce more for the increasing population and his family but save and protect the fragile environment while practicing it. Advanced technologies with appropriate and exact inputs such as precision farming can take care better of environment. Precision farming is right farming operations at right time leading to maximize profit. Every farming activity performed with precision by doing the right thing, in the right place, in the right way, at the right time and with right technology leading to maximize profit is known as precision farming. It is an efficient, eco-friendly and accurate management of agriculture inputs such as water, seed, fertilizers, pesticides and harvesting tools and machinery to maximize profit and reduce waste. The basic concept of precision farming is to match agricultural inputs and operations as per crop (variety/hybrid/cultivar) and agro-climatic conditions with accuracy to their better use efficiency. Perfection (nothing requisite is wanting) in farming is the key to precision farming.

Precision farming is based on global positioning system (GPS). The unique character of GPS is precision in time and space. Its use is spreading to all farm operations. Precision farming is an integrated information and production based farming system that is designed to increase long term, site specific and whole farm production efficiency, productivity and profitability while minimizing, waste and unintended impacts on the environment. This concept of precision agriculture is based on observing, measuring and responding to inter and intra field variability in crops.

Farm management strategy is the key to precision agriculture that uses integrated smart technologies such as sensors, GIS, GPS, satellite radars, drones, IoT (Internet of things) and data analytics to develop informed crop production and management decisions. The field variations are precisely measured with the help of technologies to optimize applications of inputs such as manures, fertilizers, pesticides, bio-agents and water use to increase crop yield at minimum cost. The aim of precision farming is to increase productivity, decrease production costs and minimize the environmental impact of farming.

Need of precision farming

Global climatic change and other negative characters of agriculture such as the decline in the total crop productivity, overexploitation of natural resources leading to their diminishing and degrading state, more or less stagnating farm incomes, absence of eco-regional approach, prevalent declining and fragmented land holdings, trade liberalization on agriculture, limited employment opportunities in non-farm sector, have become serious concerns for agricultural growth and development. Therefore, the use of precision farming technology adoption is seen potential to increase agriculture productivity. In this technology instead of managing an entire field based upon some hypothetical average condition, which may not exist anywhere in the field, a precision farming approach recognizes site-specific differences within fields and adjusts management actions accordingly.

Indian farmers are aware that their fields have variable yields across the landscape. These variations can be traced to management practices, soil properties and/or environmental characteristics. The level of knowledge of field conditions is difficult to maintain because of the large sizes and changes due to annual shifts in leasing arrangements in the farm area. So the entire farm area has to be divided into small farm units. Precision agriculture offers the potential to automate and simplify the collection and analysis of information. It allows management decisions to be made and quickly implemented on small areas within larger fields. This enables uniform attention to every spot in the small as well as large fields or orchards.

Precision agriculture is a key component of the third wave of modern agricultural revolutions- like ever green revolution and alike. The first agricultural revolution was the increase of mechanized farming 1900 to 1930. Each farmer produced enough food to feed about 26 people during this time. The 1990s prompted the green revolution with new methods of genetic modification- improved varieties and hybrids of major cereal crops, which led to each farmer feeding about 155 people. It is now expected that by 2050, the global population will reach about 9.6 billion, and food production must effectively double from current levels in order to feed every mouth. With new technological advancements in the agricultural revolution of precision farming, each farmer will be able to feed 265 people on the same acreage. This would lead to professionalism in farming with huge saving on precious inputs with enhanced productivity. The likely stress in plant growth under precision farming are minimized or eliminated. Hence the need of precision in very profession is inbuilt and agriculture or farming is no exception. Profit maximization in farming with optimum inputs is the requirement to adopt precision farming. There is need to make all possible efforts to use new technologies like precision farming to achieve evergreen revolution and sustainability.

History

In the early 1980s the concept of precision agriculture first emerged in the United States. In 1985, researchers at the University of Minnesota applied varied lime inputs in crop fields. At the time that the practice of grid sampling appeared (applying a fixed grid of one sample per hectare). Towards the end of the 1980s, this technique was used to derive the first input recommendation maps for fertilizers and pH corrections. The use of yield sensors developed from new technologies, combined with the advent of GPS receivers, has been gaining ground ever since. Today, such systems cover large acreage abroad.

In United States of America where farmers are trying to maximize profits by spending money only in areas that require fertilizer. This practice allows them to vary the rate of fertilizer across the field according to the need identified by GPS guided Grid or Zone Sampling. Fertilizer that would have been spread in areas that don't need it can be placed in areas that do need, thereby optimizing its use and saving on the quantity of fertilizer used.

Precision agriculture developed at a varying pace around the world. Major nations were the United States, Canada and Australia. In Europe, the United Kingdom was the first to go down this path, followed closely by France, where it first appeared in 1997-1998. In Latin America leading country is Argentina where it was introduced in the middle 1990s with the support of the National Agricultural Technology Institute. In fact, the development of GPS and variable-rate (VTR) spreading techniques helped to anchor precision farming management practices. In India it is being experimented at and promoted through national centres on several precision farming development centres spread over the length and breadth of the country.

Precision farming tools

Precision farming is practiced on the output several tools. Important ones are briefly explained below.

Global positioning system (GPS): It is a navigation system based on a network of satellites that helps users to record positional information (latitude, longitude and elevation) with an accuracy of between 100 and 0.01 m. It provides exact location and position on field information such as soil type, disease and insect infestation, weed appearance, water stagnation, boundaries and obstructions. Soil type, pest occurrence, weed invasion, water holes, boundaries and obstructions and others information ground /exact position is located by farmers. There is an automatic controlling system, with light or sound guiding panel (DGPS), antenna and receiver. GPS satellites broadcast signals that allow GPS receivers to calculate their position. The system allows farmers to reliably identify field locations so that inputs (seeds, fertilizers, pesticides, herbicides and irrigation water) can be applied to an individual field or spot, based on performance criteria and previous input applications.

Sensors: These are sophisticated devices that are frequently used to detect and respond to electrical or optical signals. A sensor converts the physical parameter into a signal which can be measured electrically. Sensor technologies are used to measure texture, structure, physical characters' humidity, vapours and nutrient level in the field. Sensors enable the collection of immense quantities of data without laboratory analysis. Technologies such as electromagnetic, conductivity, photo electricity and ultra sound are used to measure temperature, texture, structure, physical character, nutrient level, vegetation, humidity, vapour, air and others. With the assistance of sensors remote sensing data are used to distinguish crop species, locate stress conditions, identify pests and weeds, and monitor drought, soil and plant conditions.

Geographic information system (GIS): It is a type of computerized map which provides information on yield of crop/variety/hybrid/cultivar, soil types and testing, topography drainage (surface and subsurface), irrigation and fertilizers and pesticides application rates.

GIS is comprised of hardware, software and procedures designed to support compilation, storage, retrieval and analysis of feature attributes and location data to produce maps. GIS links information in one place so that it can be extrapolated when needed. Computerized GIS maps are different from conventional maps and contain various layers of information (e.g. soil survey maps, rainfall, crops, soil nutrient levels, pests and yield). GIS is a kind of computerized map, but its real role is using statistics and spatial methods to analyse characters and geography. A farming GIS database can provide information on filed topography, soil types, surface drainage, subsurface drainage, soil testing, irrigation, chemical application rates and crop yield. Once analysed, this information is used to understand the relationships between the various elements affecting a crop on a specific site. In addition to data storage and display, the GIS can be used to evaluate present and alternative management by combining and manipulating data layers to produce an analysis of management scenario.

Grid soil sampling and variable-rate fertilizer (VRT) application: The field is divided into several small and equal divisions. To achieve this factor is fitted with dish antenna to receive signals from satellite. Soil samples are mechanically taken from each sub division.

Soil samples are tested in a well-equipped modern soil testing laboratory for physical and chemical characteristics of soil and recorded. By the test results colour grams are prepared for entire field. The colour grams are stored in the computer for various functions. This helps in balancing soil fertility of the field. Fertilizers are then automatically applied at variable rates only to where they are needed as indicated by colour grams. This brings uniformity of soil fertility and thereby maximum economic yield, which is not possible by other methods. Variable-rate technologies (VRT) are automatic and may be applied to a number of farming operations. VRT systems set the rate of delivery of farm inputs depending on the soil type noted in a soil map. Information extrapolated from the GIS can control processes, such as seeding, fertilizer and pesticide application, herbicide selection and application at a variable rate in the right place at the right time. Grid sampling uses the same principles of soil sampling but increases the intensity of sampling. Soil samples collected in a systematic grid also have location information that allows the data to be mapped. The goal of grid soil sampling is a map of nutrient needs. We can say it an application map. Samples may be collected for more than one area of a field which fall in to the same range of yield, soil colour, etc. and thus the same zone. Grid soil samples are analysed in the laboratory, and an interpretation of crop nutrient needs is made for each soil sample. Then the fertilizer application map is plotted using the entire set of soil samples. The application map is loaded into a computer mounted on a variable-rate fertilizer spreader. The computer uses the application map and a GPS receiver to direct a product-delivery controller that changes the amount and/or kind of fertilizer product, according to the application map.

Rate controllers: They control the delivery rate of chemical inputs such as fertilizers and pesticides (liquid as well as granular). These rate controllers monitor the speed of the tractor/sprayer traveling across the field, as well as the flow rate and pressure (if liquid) of the material, making delivery adjustments in real-time to apply a target rate.

Precision irrigation in pressurized system: According to data it controls irrigation machine motions with GPS based controllers. In addition to motion control, wireless communication and sensor technologies are being developed to monitor soil and ambient conditions, along with operation parameters of the irrigation machines (i.e. flow and pressure) to achieve higher water application efficiency and utilization by the crop. Technologies and procedures have been developed for commercial use in sprinkler irrigation by controlling the irrigation machines motion with GPS based controllers. In addition to motion control, wireless communication and sensor technologies monitor soil and ambient conditions, along with operation parameters of the irrigation machines (i.e. flow and pressure) to achieve higher water application efficiency and utilization by the crop.

Software: Applying precision agriculture technologies will frequently require the use of software to carry out diverse tasks such as display-controller interfacing, information layers mapping, pre and post processing data analysis and interpretation, farm accounting of inputs per field, and many others. The most common are software to generate maps (e.g. yield, soil); software to filtering collected data; software to generate variable rate applications maps (e.g. for fertilizer, lime, chemicals); software to overlay different maps; and software to provide advanced geo-statistical characteristics. These are excellent options for precision agriculture farm management and record keeping to keep up with the needs of modern, information-intensive farming systems. Software are continuously upgraded to make them user friendly.

Yield monitors: Yield monitors are a combination of several components. They typically include several sensors and other components, including a data storage device, user interface (display and key pad), and a task computer, which controls the integration and interaction of these components. The sensors measure the mass or the volume of grain flow (grain flow sensors), separator speed, ground speed, grain. In the case of grains, yield is continuously recorded by measuring the force of the grain flow as it impacts a sensible plate in the clean grain elevator of the combine. In all yield monitors, GPS receivers are used to record the location of yield data and create yield maps. Other yield monitoring systems include devices used in forage crops to keep track of weight, moisture, and other information on a per-bale basis. Baling of forage in India in uncommon except for the supply to animal transport wings of Army.

Modern precision farming technologies

Precision agriculture is an application of breakthrough digital farming technologies. Some of the emerging technologies are:

Automation in farm machinery

The tractor does most of the work in farming. Self-steering tractors have existed for some time now, as John Deere and some others equipment works like a plane on autopilot. Technology is advancing towards driverless machinery programmed by GPS to apply fertilizer or plough land. Solar powered machine that identifies weeds and precisely kills them with a dose of herbicide or lasers are some of the other innovations.

Robot

An agbot, also called an agribot (agriculture robot), is an autonomous robot used in farming to help improve efficiency and reduce reliance on manual labour. Agriculture robots also known as AgBots, already exist, but advanced harvesting robots are being developed to identify ripe fruits, adjust to their shape and size, and carefully pluck them from branches. Future farms are expected to be tilled, sown, tended and harvested solely by fleets of co-operating autonomous robots that will remove weeds, fertilize, control pests and diseases, all the while collecting valuable data that to by remote control. Future agriculture is robotic agriculture and data based.

Drones and satellite imagery

Drone: An unmanned aerial vehicle (UAV) is commonly known as a drone. This is an aircraft without a pilot on board. UAVs are a component of an unmanned aircraft system comprising of a UAV, a ground-based controller, and a system of communications between the two. Drones can assist famers in a range of tasks from analysis and planning, to the actual planting of crops, and the subsequent monitoring of fields to ascertain health and growth. As farms become larger and more efficient to meet this escalating demand, drones will prove invaluable in precisely managing a farm's vital operations including aerial planting of trees, orchards and shrubs. Some of the major applications are listed below:

Soil and Field Analysis-Drones are able to produce precise maps for soil analysis at the beginning of the crop cycle, which helps to direct seed planting patterns. After planting, this data helps determine irrigation and nitrogen-level management.

Crop Monitoring-Drones allow real-time monitoring at more accurate and cost-effective level than previously used satellite imagery. The micro drones and accessory kit is specifically tailored to this purpose, offering users an aerial imaging package designed to monitor nutrients, moisture levels, and overall cop vigour in order to help keep crops healthy and estimate yields.

Irrigation-Drones equipped with hyper-spectral, multispectral, or thermal sensors are able to identify areas that require changes in irrigation. Once crops have started growing, these sensors are able to calculate their vegetation index, and indicator of health, by measuring the crop's heat signature.

Health Assessment-Scanning crops with visible and Infrared (IR) light, drones can identify which plants may be infected by bacteria or fungus, helping to prevent disease from spreading to other crops. With multispectral images, one can detect diseases and/or sickness before it is possible to detect it with a human eye. Farmers are able to identify and measure crop issues like disease, pest problems, weeds, and water-stress, with all data being processed into indices tailored for specific agricultural applications.

Off late drones are used for unthinkable operations like planting of trees or orchards, doing pollination in greenhouse cross pollinated crops and others.

Satellite: A satellite in general is any natural or artificial body moving around a celestial body mainly earth planet. A satellite while in the orbit performs its designated role throughout its lifetime. For example, a communication satellite is a kind of repeater station that receives signals from ground, processes them and then retransmits them back to Earth. An Earth observation satellite is a photographer that takes pictures of regions of interest during its periodic motion. These satellites are put into the desired orbit and have payloads depending upon the intended application. A weather forecasting satellite takes photographs of clouds and monitors other atmospheric parameters, thus assisting the weatherman in making timely and accurate forecasts.

Advances in drone and satellite technology benefit precision farming because drones take high quality images, while satellites capture the bigger picture. Light aircraft pilots can combine aerial photography with data from satellite records to predict future yields based on the current level of field biomass. Aggregated images can create contour maps to track where water flows, determine variable-rate seeding, and create yield maps of areas that were more or less productive. This technology is a planning tool to deliver the desired application and results.

The Internet

The internet of things (IoT) the network of physical objects outfitted with electronics that enable data collection and aggregation. IoT comes into play with the development of sensors and farm-management software. For example, farmers can spectroscopic ally measure nitrogen, phosphorus, and potassium in liquid manure which is notoriously inconsistent. They can then scan the ground to see where cows or other animals have already urinated and apply fertilizer to only the spots that need it. This cuts fertilizer uses by up to 30%. Moisture sensors in the soil determine the best times to remotely water plants. The irrigation systems can be programmed to switch which side of tree trunk they water based on the plant's need and rainfall. The need based quantitative and quality irrigation is practiced in precision farming.

As another example, monitoring technology can be used to make beekeeping more efficient. Honeybees are of significant economic value and provide a vital service to agriculture by pollinating a variety of crops, besides yielding valuable honey. Monitoring of a honeybee colony's health via wireless temperature, humidity and CO2 sensors helps to improve the productivity of bees, and to read early warnings in the data that might threaten the very survival of an entire hive. Likewise, it is of great use in several operations.

Precision farming on arable land: The use of precision farming techniques on arable land is the most widely used and most advanced amongst farmers in agriculturally advanced countries. Controlled traffic farming (CTF) is a whole farm approach that aims at avoiding unnecessary crop damage and soil compaction by heavy machinery, reducing costs imposed by standard methods. Controlled traffic methods involve confining all field vehicles to the minimal area of permanent traffic lanes with the aid of global navigation satellite system (GNSS) technology and decision support systems. Another important application of precision agriculture in arable land is to optimise the use of fertilizers, starting with the three main nutrients nitrogen, phosphorus and potassium. In conventional farming these fertilizers are applied uniformly over fields at certain times during the year. This leads to over application in some places and under-application in others. The environmental cost is directly related to over-application which allows nitrogen and phosphorus leaching from the field into ground- and surface waters or to other areas of the field where they are not desired. With the use of precision agriculture methods, fertilizers can be applied in more precise amounts, with a spatial and temporal component to optimize the application. The technology that allows the farmer to control the amount of inputs in arable lands is the Variable Rate Application (VRA), which combines a variable-rate (VR) control system with application equipment to apply inputs at a precise time and/or location to achieve site specific application rates of inputs. VRs are decided on the basis of prior measurement, e.g. from remote sensing or machine mounted sensors.

Precision farming for fruits and vegetables. In fruit and vegetable farming the recent rapid adoption of machine vision methods allows growers to grade products and to monitor food quality and safety, with automation systems recording parameters related to product quality. These include colour, size, shape, external defects, sugar content, acidity, and other internal qualities. Additionally, tracking of field operations such as chemicals sprayed and use of fertilizers can be possible to provide complete fruit and vegetable processing methods. This information can be disclosed to consumers for risk management and for food traceability as well as to producers for precision agriculture to get higher quality and larger yields with optimized inputs. New approaches have been developed that take into account not only the actual size of the tree, the condition of the crop but environmental conditions also. The precision agriculture domains with specific implications in horticultural crops captured are: data collection, yield mapping, remote sensing, quality mapping, and variable rate application. The spatial and temporal variability in orchards as well as effects of site-specific application of inputs on maximization of yields are major objectives.

Precision livestock farming. Precision livestock farming (PLF) is defined as the management of livestock production using the principles and technology from precision farming. Processes suitable for the precision livestock farming approach include animal growth, milk and egg production, detection and monitoring of diseases and aspects related to animal behaviour and the physical environment such as the thermal micro-environment and emissions of gaseous pollutants. Systems include milk monitoring to check fat and microbial levels, helping to indicate potential infections, as well as new robotic feeding systems, weighing systems, robotic cleaners, feed pushers and other aids for the stockman such as imaging systems to avoid direct contact with animals. New systems for data monitoring for feed and water consumption can be used to the early detection of infections is available now. Other developments include the monitoring on the growing herd where measurement of growth in real time is important to provide producers with feed conversion and growth rates. Sensors are now used to provide alerts concerning birthing and fertility. A vaginal thermometer monitors the temperature, imminence of birthing and the breaking of waters, and communicates to the farmer via SMS. Also, a sensor placed on an animal's collar records parameters to detect signs of oestrus and the readiness for fertilisation. An SMS message then allows the farmer to plan for insemination.

On-line resources for precision agriculture. There is a wealth of information available over the internet on new technology for farm production. Most manufacturers of farm equipment, GPS receivers, sensors, and other PA technologies use this media to inform growers on new products, technical specifications, trouble-shooting information, software upgrades, and a variety of services.

Advantages of precision farming

There are many benefits to precision farming but there are also a number of other issues which need be taken in to account before adopting the technology.

1. It enhances agriculture productivity and prevent soil degradation in farms/ cultivated fields leading to sustainable agriculture development.
2. It reduces excessive chemical usage in agriculture.
3. Global positioning system allows fields to be surveyed with ease.
4. Crop yield and soil characteristics can also be mapped.
5. Agronomic practices are followed as per specific requirement of the crop
6. Non-uniform fields can be sub-divided into smaller plots according to their specific requirements.

7. Provides opportunities for better resource management and so could reduce wastage.
8. Minimises the risk to the environment particularly with respect to nitrate leaching and groundwater contamination due to the optimisation of use of agrochemical products.
9. Dissemination of information about agriculture practices to improve quality, quantity and reduced cost of production of crops.
10. Technical perspective allows efficient time management.

Disadvantages of precision farming

1. Techniques are still under development and so it is important to take specialist advice before making expensive decisions.
2. Initial capital costs are high and so it is seen as a long-term investment.
3. It takes several years to have sufficient data to implement the system fully.
4. Extremely demanding work particularly collecting and then analysing the data.
5. All the time farmer is on the work to meet the precise demand at right time.
6. It may take several years before the actual collection of data to fully implement the system.
7. Collection and analysis of data is not an easy job on which the system works.

India's potential for precision farming

Precision farming in India has great potential due to:

1. Farmers can make use of forecast to reduce problems like water stress, diseases, insect-pests and nutrient requirement and their deficiencies.
2. It increases water use efficiencies even if water resource is poor.
3. Site specific application of crop requirements can reduce the production cost and be environment friendly.
4. Precision farming involves efficient modern farm management leading to improvement in productivity, profitability, produce quality, environment protection, on- farm quality of life, food safety and rural economic improvement.

5. It involves deployment of skill manpower and modern agriculture technology
6. It facilitates monitoring in field conditions like greenhouse farming and vertical farming.

Challenges in adopting precision farming in India

Precision farming in India could not become popular due to the following.

1. Adoption of precision farming in India is at present in nascent stage due to small land holdings with majority of farmers.
2. Lack of required infrastructure.
3. Poor risk bearing capacity of farmers.
4. Demographic conditions. Full of diversity.
5. Lack of local technical expertise.
6. Heterogeneity of cropping systems and market imperfections.
7. Pooled crop specific like sugarcane, paddy, cotton, wheat etc (where huge acreage in a block is there) precision farming with common tools possible at village level with the progressive attitude of village chief which is not there for want of awareness.

Opportunities

1. The most important component in taking up precision farming will be in creating a huge resource of engineers, scientists, skilled and trained agriculturists to develop various components of technology with the involvement of industry and research institutes. Hence capacity building on modern and sustainable agriculture, the need of India.
2. Precision farming provides an opportunity to huge industrial sector on corporate social responsibility (CSR) to invest in rural sector through hi-tech agriculture.
3. Opportunities in India- for promotion of precision farming following can help in the country;
   a. Promote precision farming to select farmers having risk bearing potential as technology requires heavy initial investment.
   b. Identify and practice in niche areas
   c. Encourage farmers to adopt water accounting protocols at farm level.

d. Evolve a policy for efficient transfer of technology to the farmers.

e. Provide complete technical back-up support to the farmers to develop pilot or models replicable on large scale.

f. Policy support on procurement prices in formulation of self-help groups or co-operatives.

g. Create zonal facilities like cold storage, processing, grading, packaging for export.

A study in Tamil Nadu on precision farming

Precision agriculture is a cropping rather than an agricultural concept. Precision farming is not a technology but management of philosophy to respond to spatial variability. Adoption of precision farming in India is poor. In Tamil Nadu in the beginning of 20th century a study on precision and non-precision farming has revealed that adoption of precision farming leads to about 80 per cent increase in yield in tomato and 34 per cent in brinjal/eggplant. Increase in gross margin has been found 165 per cent and 67 per cent in tomato and brinjal production, respectively. The contribution of technology for higher yield in precision farming has been recorded as 33.7 per cent and 20.5 per cent, respectively in tomato and brinjal. The elasticity of 0.39 for the adoption in tomato and of 0.28 in brinjal indicated that as the probability of adoption increases by 10 per cent, the net return increases by 39 per cent and 28 per cent in tomato and brinjal cultivation, respectively. Lack of finance and credit facilities have been identified as the major constraints for non-adoption of precision farming. The study has suggested that providing of subsidies for water soluble fertilizers and pump-sets can increase the adoption of precision farming.

Precision Farming Development Centres (PFDC) in India

To promote precision farming Ministry of Agriculture and Farmers' Welfare is experimenting and demonstrating precision farming in different parts of the country. At present (2019) there are 16 Precision Farming Development centres (PFDC) in India located at Jorhat (Assam), Samastipur (Bihar), Hisar (Haryana), Bikaner (Rajasthan), Raipur (Chhattisgarh), Hyderabad (Telangana), IARI, Delhi, KAU (Kerala), Bhubaneshwar (Odisha), Bengaluru (Karnataka), Rahuri (Maharashtra), Navsari (Gujarat), Lucknow (Uttar Pradesh), Pantnagar (Uttarakhand), Coimbatore (Tamil Nadu) and IIT Kharagpur (West Bengal). The demonstrations of precision agriculture befitting to the geographic region are carried out along with development of innovations.

2

Laser Land Levelling (LLL)

It is a common knowledge that most of the farmers apply irrigation water until all the points in the field are fully wetted and covered with a thin sheet of water. Studies have indicated that a significant (20-25%) amount of irrigation water is lost during its application at the farm due to poor farm designing and unevenness of the fields. This problem is more pronounced in the case of rice fields. Unevenness of fields leads to inefficient use of irrigation water and also delays tillage and crop establishment options. Fields that are not levelled have uneven crop stands, increased weed burdens and uneven maturing of crops. All these factors tend to contribute to reduced yield and produce quality which reduce the potential farm income. Effective land levelling is meant to optimize water-use efficiency, improve crop establishment, reduce the irrigation time and effort required to manage crop.

Reasonably levelled agriculture fields has advantage over uneven and undulating crop fields as unevenness of the soil surface has a significant impact on the seed germination, crop stand and its yield. Traditionally land levelling was being done throughout India by bullock drawn plough or tractors and applying different devices for levelling the field by visual assessment e.g. on small plots with hoes, or with draft animals and equipment such as ploughs and bars or scrapers or grading. Equipment such as grading blades and hydraulically operated levellers mounted on wheels are used with tractors. More advanced and sophisticated levelling equipment is operated with a laser emitter, a laser sensor or receiver, and a scraper pulled by a tractor. After the desired level or slope of the field and/or the difference of the high and low spots have been surveyed, the emitter is set to send a rotating laser beam creating a plane of laser light above the field surface. The laser light is used as the levelling reference. It directs the hydraulic system of the moving tractor and scraper, and thereby controls the levelling. Hence, laser levelling or laser land levelling is levelling the crop field within certain degree of desired slope using a guided laser beam throughout the field.

Importance

The importance of land levelling by laser is the accuracy in levelling the crop fields to realise the following main benefits at crop field.

- Improved crop establishment.
- Even water coverage of the field and saving in water with increased water use efficiency.
- Even crop stands and uniform maturation.
- Reduction of weeds making weed removal/control cheaper.
- Increase of farming area by 5-7 %.
- Reduction of farm operation times by 10-15 %.
- Within rice production, the possibility of changing from planting and transplanting to direct seeding which results in reduced labour requirement.
- Average crop yield increase of 10- 20 %.

With above and many more advantages laser land levelling has become popular among Indian farmers. Small farmers can also afford to rent laser levellers, enabling more efficient use of water and boosting food production. Laser levelling has resulted in uniform and higher production in a field irrespective of its size.

Principle of laser land levelling

Laser land levelling system includes three parts: laser transmitter and receiver, control system and scraper. In fact, it is like a scraper with rear wheels, which could automatically up and down according to the height of the ground. The scraper goes up and down through the telescopic oil cylinder. The cylinder is installed between the rear axle and scraper. The process of moving forward, when ground is higher, the oil cylinder would be shortened to make the scraper down, then scraper blade cut soil there and goes forward with the soil. When the ground is lower, the oil cylinder will be extended to make oil scraper up, and then the soil in the scraper will be filled here. During the land levelling, soil in the higher places will be scraped to flatten the lower places to complete the land levelling.

Laser land levelling technique

Laser levelling is a process of smoothing the land surface (plus minus 2cm) from its average elevation using laser equipped drag buckets to achieve precision in land levelling. Precision land levelling involves altering the field in such a way as to create a constant slope of 0 to 0.02 %. This practice makes use of large horsepower tractors and soil movers that are equipped with GPS and laser guided instrumentation so the soil can be moved either by cutting or filling to create the desired slope/level.

Components of laser leveller

Drag bucket

Laser transmitter

Laser receiver

Control panel

Hydraulic control system

The process of land levelling

1. Set the average elevation value of the field in control box.
2. Laser controlled bucket be positioned at a point that represent the mean height of the field.
3. The cutting blade should be set slightly above ground level (1.0-2.0 cm)
4. The tractor then be driven in a circular direction from the high areas of the field to the lower areas.
5. As soon as the bucket is near filled with soil the operator should turn and drive the tractor towards the lower area. Similarly, as soon as the bucket is near empty the tractor should turn and driven back to the higher area.
6. As soon as the whole field is covered in the circular manner, the tractor with bucket then should do a final levelling pass in long runs from high end of the field to the lower end.
7. The field then be re-surveyed to make sure that the desired level of precision has been achieved.

Laser land levelling reduces the use of water for irrigation and increases crop yields. Overall, adoption of laser land levelling technology promotes agricultural productivity, energy use efficiency and water use sustainability.

3

Mechanized Direct Seed Sowing

Sowing of Seed: It is a process of planting seeds of crops in well prepared or no till field or nursery beds/pots for raising crops in following methods:

Methods of Sowing: The sowing method is determined by the crop to be sown and resources on the disposal of farmer. There are 6 sowing methods which differ in their merits, demerits and adoption. Those are:

1. Broad casting
2. Broad or Line sowing
3. Dibbling
4. Transplanting
5. Planting
6. Putting seeds behind the plough.

1. **Broad casting:** It is the scattering of seeds by hand all over the prepared field followed by covering with wooden plank or harrow for contact of seed with soil. Crops like wheat, paddy, sesame, fenugreek, coriander, and others. are sown by this method.
2. **Drilling or Line sowing:** It is the dropping of seeds into the soil with the help of implement such as mogha, seed drill, seed-cum-fertilizer driller or mechanical seed drill and then the seeds are covered by wooden plank or harrow to have contact between seed and soil. Crops like jowar, wheat bajara, etc. are sown by this method.
3. **Dibbling:** It is the placing or dibbling of seeds at cross marks (+) made in the field with the help of maker as per the requirement of the crop in both the directions. It is done manually by dibbler. This method is followed in crops like groundnut, castor, and hy. cotton, and others which are having bold size seed and high value.
4. **Transplanting:** It is the raising of seedlings on nursery beds or house and transplanting of seedlings in the laid out field. For this, seedlings are

allowed to grow on nursery beds for about 3-4 weeks. Beds are watered one day before the transplanting of nursery to prevent jerk to the roots. The field is irrigated before actual transplanting to get the seedlings established early and quickly which reduce the mortality. Besides the advantages and disadvantages of dibbling method, initial cost of cultivation of crop can be saved but requires due care in the nursery. This method is followed in many crops like paddy, fruit, vegetable, crops, tobacco, and others.

5. **Planting:** It is the placing of vegetative part of crops which are vegetatively propagated in the well laid out field. For example, seed tubers of potato, mother sets of ginger and turmeric, cuttings of sweet potato and grapes, sets of sugarcane and others.

6. **Putting seeds behind the plough:** It is dropping of seeds behind the plough in the furrow with the help of manual labour by hand. This method is followed for crops like gram in some areas for better utilization of soil moisture. The seeds are covered by successive furrow opened by the plough/tractor tine. This method is commonly followed for sowing of the crops like wheat.

Mechanized direct seed sowing

To perform above operation with accuracy and in short time machines or seed drills operated by prime mover are used. Recently, the use of rice direct seeding has been increasing rapidly owing to rural labour shortages and continuous increases in agricultural production costs.

Seed drill:A seed drill is a device that sows the seeds for crops by metering out the individual seeds, positioning them in the soil, and covering them to a certain desired average depth. This makes sure the seed will be placed evenly. The seed drill sows the seeds at equal distances and proper depth, ensuring that the seeds get covered with soil and are saved from being eaten by birds and being blown by the wind. This allows plants to get sufficient sunlight, nutrients, and water from the soil. The mechanized seeding is known as drilling or drilling seeds.

Traditionally, a seed drill used to consist of a hopper filled with seeds arranged above a series of tubes that can be set at selected distances from each other to allow optimum growth of the resulting plants. Seeds are spaced out using fluted paddles which rotate using a geared drive from one of the drill's land wheels—seed rate is altered by changing gear ratios. Most modern drills use air to convey seed in plastic tubes from the seed hopper to the coulters—it is an arrangement which allows seed drills to be much wider than the seed hopper—as much as

12 m wide in some cases. The seed is metered mechanically into an air stream created by a hydraulically powered on board fan and conveyed initially to a distribution head which sub-divides the seed into the pipes taking the seed to the individual coulters.

The seed drill allows farmers to sow seeds in well-spaced rows at specific depths at a specific seed rate; each tube creates a hole of a specific depth, drops in one or more seeds, and covers it over. This invention gives farmers much greater control over the depth that the seed is planted and the ability to cover the seeds without back-tracking.

The use of a seed drill also facilitates weed control. Broadcast seeding results in a random array of growing crops, making it difficult to control weeds using any method other than hand weeding. A field planted using a seed drill is much more uniform, typically in rows, allowing weeding with the hoe during the growing season. Weeding by hand is laborious and inefficient. Poor weeding reduces crop yield, so this benefit is extremely significant.

Before the operation of the seed drill, the ground must be ploughed and harrowed. The plough digs up the earth and the harrow smooths the soil and breaks up any clumps. The drill must then be set for the size of the seed used. Afterwards, the grain is put in the hopper on top which then follows along behind the drill while it spaces and plants the seed. This system is still used today but has been modified and updated such that a farmer can plant many rows of seed at the same time.

A seed drill can be pulled across the field using bullocks or a tractor. Seeds sown using a seed drill are distributed evenly and placed at the correct depth in the soil.

Zero till drill or seed drill: It is direct seeding without any mechanical disturbance of soil. It ensures early sowing of wheat and reduces weed infestation particularly *Phalaris minor* in wheat. After harvesting paddy in northern India this practice is popular for sowing wheat. This technology ensures at least 30 percent crop residue incorporation in soil, which results in increased organic matter content, improved retention and transmission characteristics of soil and improves soil aggregation and structure.

Zero tillage technology, also known as use of happy seed drill technology, is also effective in soil water conservation and increases water-use-efficiency of the crops. It maintains desirably high and economic level of productivity, cut short the need for chemical amendments and pesticides, preserve ecological stability and minimizes the pollution of natural water and environment. Drill technology reduces input cost by reducing fuel consumption and increases the net profit of farmers. The technology has become popular in Indo-Gangetic plains of India where rice – wheat cropping system is prevalent.

Advantages

1. It saves fuel and time hence reduction in production cost.
2. Good germination is obtained as the moisture needed for seed germination is adequate
3. There is proper distribution of seeds and fertilisers.
4. In adds organic matter to the soil and improves soil health.
5. It is environment friendly.

4

Seedling and Sapling Transplanting

Seedling:A seedling is a very young plant that grows from a seed. When the moisture, light, and temperature conditions are correct, the seedling's development begins with seed germination and the formation of three main parts:

1. Radicle - Embryonic root
2. Hypocotyl - Embryonic shoot
3. Cotyledons - Seed leaves.

In vegetables and annual ornamental crops where generally seed size is small they are germinated in nursery beds or in plug trays indoor (greenhouse) or outdoor(field), hardened before transplanting.

Sapling: A sapling is a young tree, say more than 1 year and less than 3 years' old. In the life cycle of trees, a seedling converts into a sapling as the next stage of development. It is nearly one-meter tall and generally remains to be known as a sampling until its stem reaches 7 centimetre in diameter. The time period of this stage depends upon the particular species of a tree. In vegetable and annual ornamental crops seedlings generally are one month to 40 days old. Older than this, seedlings do not give better yields on transplanting. Very young seedlings exhibit larger mortality on transplanting.

Transplanting: Transplanting or replanting is the technique of moving a plant by uplifting from one location to another. From nursery beds seedlings or saplings are shifted and planted either in the field or greenhouses for commercial growing. Seedlings are uprooted if grown in soil medium. They are transplanted with growing medium. Uprooted seedlings take time to establish at a new place and face transplanting shock- time taken by roots to establish contact with the soil or growing medium.

Steps to transplant

1. Seedlings should be hardened-off, well-fed and watered before transplanting.

2. Prepare a weed-free surface. Loosen and aerate soil by tilling or hoeing.
3. Dig/make a hole large enough for seedling.
4. Carefully remove seedling from its container/ or bundle of uprooted seedlings. Try not to disturb the roots.
5. Set seedling in hole level with soil surface. The exception is tomato seedlings, which can be transplanted a bit deeper.
6. Feed seedling to kick start growth. Transplant each seedling with a hefty handful of compost. If you don't make compost, purchase specially formulated fertilizer for transplanting.
7. Surround seedling with displaced soil.
8. Water seedling thoroughly, avoid over watering.
9. Mulch soil before transplanting to maintain soil moisture and regulate temperature.
10. Keep area weed-free.

Note: Plastic mulching and laying of micro-irrigation(drip) is generally done prior to transplanting.

For successful transplanting

Start with healthy seedlings: If you start your own seeds, make sure your seedlings have plenty of light, adequate water and drainage. A good growing medium is critical for root establishment. If you buy seedlings remember biggest doesn't always mean best. Look for healthy and consistent leaf colour. The roots should be deep, long, white and fibrous. The stem should be thick and strong. The growing medium should be held together by tight tangled roots. Do not choose leggy seedlings; too long of stem means the seedling was starved for light. Seedling exhibiting yellowing of seedlings indicate starved seedling mainly of nutrients.

Hardening off is important: Don't skip this step. Hardened seedlings should be purchased. Seedlings need to be acclimated to the natural environment of the field before transplanting. If you have raised seedlings in green house harden them for about a week under environment similar to the field. In winter put seedlings in sun and bring them in at night.

Timing is everything: Transplanting is generally done in the evening hours to get better establishment. Overcast days are good days to transplant. Saplings, generally in fruit crops are planted during monsoon months after preparing pits and putting adequate organic manures in the pits along with recommended amount of fertilizers. The planting distance is specific to the fruit crop varieties or hybrids or cultivars. After transplanting watering is a must which should be appropriate avoiding over and under watering.

Precision in depth and spacing of seedling

For precision planting and transplanting, crop wise typical farming system that delivers data-driven planting advice is followed which determines variable planting rates ensuring precise depth and spacing to accommodate varying conditions across a single field, in order to maximize yield. The depth and spacing of seedling affects their establishment in soil avoiding competition for water and nutrients. There are mechanical transplanters having for different crops particularly tray plugs seedlings in vegetables and other crops.

5

Soil Culture

In soil culture, also known as in-ground culture, crops are raised on level ground as well as in raised beds. Soil culture is more often used in growing vegetables (e.g., tomatoes, lettuce, cucumbers, and peppers) rather than ornamental crops such as herbs, roses, freesia, and foliage plants as well as popular with smaller commercial and organic growers. The soil should be well drained, fertile, have a medium to light texture and be high in organic matter. The soil in a greenhouse should be worked as deeply as possible to break any existing hardpans or those that may have developed because of repeated traffic over the area. Fumigation or pasteurization of the soil is a must before each crop or at least once a year. This will help to destroy disease, nematodes and weeds. If soil is steamed, hold it at 180 degrees F (82°C) for 4 hours. Production techniques used in soil culture are similar to intensive crop production methods used in the field. Although soil is readily available at most places where a greenhouse is located most commercial growers, no longer rely on soil culture in the greenhouse as they did ealier. Some of the reasons include insect and disease pressures and costly fumigation.

Compost or manure addition

Field soil in a greenhouse is usually amended with a large volume of organic matter such as well-made compost, farm yard manure, vemi-compost before growing greenhouse vegetables and other crops. Apply the compost or manure before the soil is fumigated so weed and disease pests in the compost and manure are destroyed. Undecomposed manure should be avoided or apply at least 4 months prior to harvest to avoid contamination from human pathogens such as *E. coli* and others. Greenhouse soils amended with compost or manure or soil amendments may or may not require additional fertilization or liming for good plant and root growth. Soil analysis and guidance of agronomist would give better performance of the crop. For organic production under greenhouses organic standards or appropriate recommendations on the subject would provide better results.

Growing Media

Growing medium is perhaps the most important used in greenhouses. Performance of crop is obviously dependent on growing medium and its enrichment. Growing media must be amended to provide the appropriate physical and chemical properties necessary for plant growth. Field soils are generally unsatisfactory for the production of plants in containers. This is primarily because soils do not provide the aeration, drainage and required water holding capacity. Several soilless growing media have also been developed. The following are used for amendment of growing medium.

Peat and Peat-Like Materials

Peat moss is formed by the accumulation of plant materials in poorly drained areas. The type of plant material and degree of decomposition largely determine its value for use in a growing medium. Although the composition of different peat deposits vary widely, four distinct categories may be identified:

Hypnaceous moss – this type of peat consists of the partially decomposed remains of hyprum, polytrichum and other mosses of the Hypanaceae family. Although it decomposes more rapidly than some other peat types, it is suitable for media use.

Reed and Sedge – are peats derived from the moderately decomposed remains of rushes, coarse grasses, sedges, reeds and similar plants. These fine textured materials are generally less acid and contain relatively few fibrous particles. The rapid rate of decomposition, fine particle size and insufficient fiber content make reed and sedge peats unsatisfactory for media use.

Humus or Muck – consists of the decomposed debris of finely divided plant materials of unknown origin. Humus often contains large quantities of silt and clay particles, and when mixed with soil does not improve drainage or aeration. Due to its rapid rate of decomposition and particle size, humus is considered to be undesirable for growing media use.

Sphagnum moss – is the dehydrated remains of acid-bog plants from the genus Sphagnum (i.e. Spapillosum). It is light in weight and has the ability to absorb 10 to 20 times its weight in water. This is attributed to the large groups of water holding cells, characteristic of the genus. Sphagnum moss contains specific fungistatic substances which accounts for its ability to inhibit damping-off of seedlings.

Sphagnum moss is perhaps the most desirable form of organic matter for the preparation of growing media. Drainage and aeration are improved in heavier soils while moisture and nutrient retention are increased in lighter soils.

Wood Residues

Wood residues constitute a significant source of soilless growing media. These materials are generally bi-products of the lumber industry and are readily available in large quantities. Nitrogen depletion by soil microorganisms, during the decomposition process, is one of the primary problems associated with these materials. However, supplemental applications of N to the growing media can make most wood residues valuable amendments.

Leaf Mold – Mango and other fruits as well as other trees are among the principle leaf types suitable for the preparation of leaf mold. Layers of leaves and soil are composted together with small amounts of nitrogenous compounds for approximately 12 to 18 months. The use of leaf mold can effectively improve the aeration, drainage and water holding properties of a growing media. Although these materials are readily available at low cost, leaf mold is not extensively used in container production.

Sawdust – the species of tree from which sawdust is derived largely determines its quality and value for use in a growing media. Several sawdusts, such as walnut and non-composted redwood, are known to have direct phytotoxic effects. However, the C:N of sawdust is such that it is not readily decomposed. The high cellulose and lignin content along with insufficient N supplies creates depletion problems which can severely restrict plant growth. However supplemental applications of nitrogen can reduce this problem.

Barks – are primarily a bi-product of the pulp, paper and plywood industries. Suitable particle size is obtained by hammer milling and screening. This produces a material which is suitable for use in container media. Physical properties obtained from tree barks are similar to those of Sphagnum moss.

Bagasse

Bagasse is a waste bi-product of the sugar industry. It may be shredded and/or composted to produce a material which can increase the aeration and drainage properties of container media. Because of its high sugar content, rapid microbial activity results after the incorporation of bagasse into a media. This decreases the durability and longevity of bagasse and influences N levels. Although bagasse is readily available at low cost, (usually transportation), its use is limited.

Rice Hulls

Rice hulls are a bi-product of the rice milling industry. Although they are extremely light in weight, rice hulls are very effective at improving drainage. The particle size and resistance to decomposition of rice hulls and sawdust are very similar. However, N depletion is not as serious of a problem in media amended with rice hulls.

Several other organic materials are suitable for use with container media. Included are: manures; corn cobs; straw; peanut and pecan shells. However these do not constitute major commercial sources of organic amendments.

Sand

Sand, a basic component of soil, ranges in particle size from 0.05mm to 2.0mm in diameter. Fine sands (0.05mm – 0.25mm) do little to improve the physical properties of a growing media and may result in reduced drainage and aeration. Medium and coarse sand particles are those which provide optimum adjustments in media texture. Although sand is generally the least expensive of all inorganic amendments it is also the heaviest. This may result in prohibitive transportation costs. Sand is a valuable amendment for both potting and propagation media.

Perlite

Perlite is a silicous mineral of volcanic origin. The grades used in container media are first crushed and then heated until the vaporization of combined water expands it to a light powdery substance. Lightness and uniformity make perlite very useful for increasing aeration and drainage.

Perlite is very dusty when dry and has a tendency to float to the top of a container during irrigation. It has also been shown that perlite contains potentially toxic levels of fluorine. Although costs are moderate, perlite is an effective amendment for growing media.

Vermiculite

Vermiculite is a micacious mineral produced by heating to approximately 745°C. The expanded, plate-like particles which are formed have a very high water holding capacity and aid in aeration and drainage. Vermiculite has excellent exchange and buffering capacities as well as the ability to supply potassium and magnesium. Although vermiculite is less durable than sand and perlite, its chemical and physical properties are very desirable for container media.

Calcined Clays

Calcined clays are formed by heating montmorrillonitic clay minerals to aproximately 690°C. The pottery-like particles formed are six times as heavy as perlite. Calcined clays have a relatively high cation exchange as well as water holding capacity. This material is a very durable and useful amendment.

These inorganic soil amendments are generally utilized to increase the number of large pores, decrease water holding capacity and improve drainage and aeration. Other materials such as: pumice; cinders; and pea-gravel are also suitable for this use.

Several synthetic soil amendments are bi-products of various plastic manufacturing companies. Others are designed specifically for use in container media. These materials are frequently used in place of sand and perlite and have much the same influence on media properties.

Expanded Polystyrene

Polystyrene flakes, a bi-product of polystyrene processing, are highly resistant to decomposition, increase aeration and drainage, and decrease bulk density. Polystyrene may be broken down by high temperatures and by certain chemical disinfecting agents.

Urea Formaldehydes

This material is prepared by mixing air with a liquid resin and allowing to cool. Urea formaldehyde foams have a greater water holding capacity than polystyrene but are similar in their influence on aeration and drainage. Raw materials are easily transported and are very effective amendments.

Preparing Soilless Growing Media

The qualtity of amendments may vary but the growing medium must have most of the following characteristics.

1. porus and well drained, yet retentive of sufficient moisture to meet the water requirements of plants between irrigations;
2. relatively low in soluble salts, but with an adequate exchange capacity to retain and supply the elements necessary for plant growth;
3. standardized and uniform with each batch to permit the use of standardized fertilization and irrigation programs for each successive crop;
4. free from harmful soil pests; pathogenic organisms, soil insects, nematodes and weed seeds;

5. biologically and chemically stable following pasteurization; primarily free from organic matter that releases ammonia when it is subjected to heat or chemical treatments.

Since innumerable amendment combinations can produce a growing medium with these characteristics, it is important to consider both the economic as well as cultural optimums. Factors that determine the cost of a growing medium include: transportation, labor, equipment, materials and handling. In many cases the cost of mixing a "custom" growing medium exceeds that of the commercially prepared materials. These factors should be studied carefully before making a decision.

Growing media in use

The composition of a growing medium is determined by the crop to be grown. There are certain common components also in the medium irrespective of the crop to be taken. The most commonly used soilless mixtures are as follows.

Table: Soilless mixtures for greenhouse crops commonly used.

Ratio/Volume	Contents
2:1	Peat, Sand
2;1	Peat, Perlite
3:1	Peat, sand
2:1:1	Peat, Perlite, Vermiculite
2:1:1	Peat , Bark, Sand
2:1:1	Peat, Bark, perlite
3;1:1	Peat, Bark, Sand
3:1:1	Peat, Perlite, Bark

Gravel Culture

Gravel culture is a general term which applies to the growing of plants without soil in an inert medium into which nutrient solutions are usually pumped automatically at regular intervals. Haydite (shale and clay fused at high temperatures), soft- or hard-coal cinders, limestone chips, calcareous gravel, silica gravel, crushed granite and other inert and slowly decomposing materials are included in the term "gravel".

Coco peat: Most commonly used medium for soilless cultivation. Coco peat is a natural fibre made out of coconut husks. The extraction of the coconut fibre from husks gives us this by-product called coco peat. It is a 100% natural growing medium. This coco peat dried in the natural sun, are processed to produce different items namely coco peat block, coco peat briquettes, coco peat tablets etc. This makes an excellent growing medium for hydrophonics or container plant growing. Clean coir is reported to have natural rooting hormones and anti-fungal properties.

Coco peat

Some of the uses of coco peat are as under:

- Coco peat is resistant to bacterial and fungal growth.
- Unlike peat coco peat holds water rather than shedding it.
- It holds water 8-9 times of its weight.
- It has the ability to store and release nutrient to plants for extended periods of time.
- It also has great oxygenation properties important for healthy root development.
- The pH of coco peat is 5.2-6.8 which is neutral to slightly acidic. This makes it great for alkaline horticulture soils.
- It is easy to use and light in weight.
- It is reusable up to 4 years.
- It makes good bedding material for earthworm which are useful in agriculture.
- Coco peat disintegrates slowly and takes years to disintegrate.
- Coco peat can be used as stand-alone or mixture with other ingredients like vermi-compost.

It is an affordable input to greenhouse farming particularly raising crop seedlings

Potting mixture: The common medium for growing plants, is prepared by mixing soil, sand and farm yard manure (compost) in 1:1:1 proportion. Coco-peat may be added to an extent of 20 to 50 % by volume to the potting mixture. It can be used for both indoor and outdoor purposes and the pH ranges from 5.5 to 6.5.

The pot mix must be sufficiently firm to hold the plant during rooting and supply food and water for the successful growth of plant. The ingredients used in potting soil are peat sand, perlite and compost made of several type of biomass. In India commercial pot mix is not commonly available. One has to prepare his own based on the ingredients are easily available.

Some plants require pot soil mix that is specific for their environment.

Common pot mix

Pot mix or soil is generally a garden soil mixed with sand and compost in a ratio of 3:1:1. This mix is readily and easily available but not desirable due to heavy weight and may be infested with pathogen, insects, weed seeds despite pasteurization or fumigation.

Desirable potting mix

Soil being a source of pathogen and weed seeds as well as heaviness nature be avoided in pot mix. Pot mix should be light, easily transportable, packageable, contain nutrient or enriched periodically with nutrients through irrigation water or separately, not compact but airy. Coco peat with and without other mix other than soil is considered better growing mix in nurseries and greenhouses in India. For greenhouses soils in raised beds are mostly having garden loam soil, sand and compost in a ratio of 1:1:1. This is to be analysed of its nutrient contents and supplemented as crop/variety/hybrid nutrient and water requirement maintaining desirable pH and EC.

6

Type of Soil Required

Crops are grown in soil. There are different types of soil. The soil type helps farmers decide what crops will grow best because some crops do better with particular types of soil. Testing the soil helps determine what nutrients are available or unavailable for growing certain crops. Farmers usually have local agricultural experts test the soil on their land for information on how to proceed in growing crops. The best soil has a balance of nutrients, such as potassium, magnesium, and others. Certain crops may need more of one nutrient, so farmers put back those ingredients into the soil for better crops.

Six different types of soil are listed below along with the crops which do well in these soils.

1. **Loam:** This type of soil helps grow the best possible crops because it provides the necessary elements for plant growth and development. Loamy sand contains mostly sand plus clay and silt. Such soils are good for root vegetables, such as carrots, radish, turnip, beets, leafy vegetables, potatoes and tomatoes.
2. **Clay**: Cabbage and broccoli grow well in clay soil, but it is not good for root vegetables because of its dense texture which does not allow free flow of water.
3. **Sand:** Root vegetables, such as turnips, parsnips, and carrots, grow well in sandy soil.
4. **Silt**: Silty soil is fertile and offers nutrients for plant development. Lettuce, cabbage, carrots, turnips, and many other vegetables flourish in silt. Strawberries, raspberries, blackberries do well.
5. **Peat**: Peaty soil provides plant growth when mixed with rich organic matter, lime, and compost that reduces its acidity. Legumes, root crops, cabbage, and spinach are easily grown in soil that's well-drained.
6. **Chalk**: Sweet corn, beets, spinach, and cabbage are among the vegetables that benefit from chalkier soil. It is recommended that farmers use fertilizers and balance the pH levels to deal with the alkaline that can sometimes stunt growth.

Farmers must get the greenhouse soils tested and get guidance what to grow and how to grow in these soils. Sandy loam soils are best choice for greenhouses.

In greenhouses soil serves as growth medium for plants. Sandy loam soil is the best choice. It should be supplemented with well-rotten FYM, compost, or bio-fertilizers. Inorganic fertilizers need to be used as per recommendations. Continuous cultivation inside the greenhouse may give rise to build up of pathogens. To destroy pathogens, pasteurization of soil is recommended (steam aeration) before every crop and soil replacement is recommenced after every 3 years. Steam aeration temperature of 60°C for 30 min is good enough to accomplish the task. Chemical fumigation can also be used but steam is the best as it is not selective thus attacking all sort of pathogens. Soil degradation is the universal problem in greenhouse soil. It affects the sustainable use of greenhouse soil and also endangers production potential.

Soil degradation in greenhouses

Soil salinity

Greenhouse cultivation often requires a greater degree of management and larger input of nutrients and irrigation. However, over-fertilization and over-irrigation result in nutrient accumulation in soil and led to soil salinity. Salinity reduces phosphate uptake and accumulation in crops grown in soils primarily by reducing phosphate availability. Salinity can directly affect nutrient uptake, such as Na^+ reducing K^+ uptake or by Cl^- reducing NO_3^- uptake. Salinity can also cause a combination of complex interactions that affect plant metabolism, susceptibility to injury or internal nutrient requirement.

Continuous cropping

Continuous cropping obstacle has become a main problem for sustainable production of vegetable, for example, serious plant diseases and fall of yield, which intimidated facility vegetables production sustainable development. As the development of specialized vegetable production, continuous cropping is inevitable in greenhouses. A lot of vegetable such as tomato, pepper, eggplants, and cucumber suffer continuous cropping obstacle. At the same time, poorly differentiated plant mass getting into the soil under continuous cropping negatively affects soil organism communities. This negative action leads to the change in metabolism of microbes from the primary to the secondary and lowers crop production

Soil acidification

A number of agricultural practices have expanded the areas of acidification soils. The main causal factor is the growth of plants that use large amounts of basic ions (e.g. legumes); particularly when fertilizers that leaves acidic residues (such as Superphosphate) are used.

Nutrition accumulation

At present, fertilizer application rates in intensive agricultural systems have increased dramatically in recent years, especially in greenhouse production systems. These high fertilizer inputs and the extremely low crop recoveries of fertilizer nutrients lead to marked deterioration in soil and groundwater quality and the systems become unsustainable. Soil chemical and biological properties in greenhouse vegetable production may change dramatically after several years of continually high inputs of fertilizers and irrigation water and planting of shallow rooting crops compared with open field cereal production systems.

Compaction

Some researcher found that bulk density was higher in greenhouse soils than uncovered soil. In recent years, there has been increasing interests in intensive vegetable cultivation, this caused the soil compacted. Compaction of soils causes a reduction in soil pore space. This reduces the rate at which water can infiltrate and drain through the soil. It also reduces the available space for oxygen in the plant root zones. For this reason, some of the major consequences of compaction are poor drainage, poor aeration, and hard pan surfaces which cause runoff. Repeated cultivation of some soils leads to a breakdown of soil structure and this also increases the likelihood of compaction.

Chemical residue

Although not as large a problem as some of the other types of soil degradation, the presence of chemical residues can be quite a problem on a local scale. These residues derive almost entirely from long term accumulation after repeated use of pesticides, etc., or of use of pesticides or other chemicals with long residual effects. Some problems that result from chemical residues include toxic effects on crop species and contamination of workers, livestock and adjacent streams.

For prevention of soil degradation in greenhouses follow crop rotation having essentially legume crop after non-legume. Use different cropping rotation models for to build up useful microbes in the soil. Add organic matter to greenhouse soils as per soil test.

7

Mapping of Soils and Plant Attributes

Soil map is a map showing diversity of soil types and its properties such as pH, texture, organic matter and depth of horizons in the area of interest obtained as the end of soil survey. Generally, soil maps are used to simply identify soils and their properties, but are sometimes required for more specific purposes, such as the suitability of soils for a particular crop, suitability for irrigation, erosion risk and many other specific needs or environmental threats.

Soils fulfil many functions important for agriculture, forestry and the management of soil resources and natural hazards. The functionality of soils depends on their properties; hence, accurate and spatially highly resolved maps of basic soil properties such as texture, organic carbon content and pH for specific soil depth are needed for the sustainable management of soils. The surface soil organic carbon (SOC) concentration is a useful soil property to map soils, interpret soil properties, and guide fertilizer and agricultural chemical applications.

Traditionally farmers apply fertilizers, chemicals and other crop production inputs to optimize the production of the field as a whole. This results in application of over application of inputs and low inputs in certain locations of the field because of variation in field characteristics including soil organic carbon, soil texture, soil nutrients, field topography, and other properties. In addition, uniform applications may increase the chances of pollution of the environment due to excess application in some field area. Precision farming technology has been shown to optimize application rates if the variation of field characteristics can be used to guide the application rate of crop-production inputs. The organic carbon concentrations of surface soil have been used to spatially vary the application rate of some crop-production inputs. The surface soil organic carbon concentration affects the activity of many herbicides, influences plant-available nitrogen and also affects the soil's ability to adsorb plant nutrients. Knowing its concentration may therefore be useful, especially if its spatial distribution could be determined accurately and for low cost.

Soils are mapped with a specific aim in mind, such as the suitability of soils for a particular crop, suitability for irrigation, erosion risk and many other specific needs or environmental threats. Most organised soil surveys in the past have been general purpose surveys. These have the advantage that they provide the basis for many different uses, some of which may yet not be known.

A Mapping Soils

Materials

- base map
- soil auger
- field kits
- clinometer

Steps

1) Visually evaluate features of the parcel and recognize the summit, back slope, foot slope, and toe slope and relate them to your base map (aerial photo).
2) Considering topographic position, landscape signatures of edaphic properties (e.g. vegetation changes or surface wetness) and your concept of possible soil distribution along a transect from the summit to the toe slope. Delineate – in pencil – probable boundaries between soil types (no more than 6 total map units).
3) To develop an understanding of how soils vary on this parcel, conduct a sampling transect starting at the top of the hillslope, working towards the creek, both within possible soil map units and near the probable boundaries between and with your delineated units. At each sample point, record the parent material, horizon information and the depth to redox features (if present). Mark the approximate sample location on your map. This transect will be completed by the entire class (ie you won't be expected to complete the entire transect) and the data will be collected will shared with everyone. We (collectively) won't take more than 3 samples within a map units or 1 to 2 near boundaries between soil types. Through these samples you are trying to ascertain the uniformity of the soil(s) within the unit and their boundaries.
4) Ascertain the slope class for each division in the parcel by using a clinometer.
5) At each sampling location use the soil auger to characterize each soil to a depth of at least 24 inches (or presence of a restrictive layer) with a specific reference to parent material, depth of A horizon, colour of A

horizon, depth to redox features, presence or absence or a fragipan, textural class of the soil fines, and pH of the B horizon. Also record the dominate vegetation types and mark the approximate position of the sampling locations on your map.

6) Evaluate the sample date within your delineations see exercise B below. Determine what probably soil it is and document how it may vary within your delineation. Correct your boundaries to reflect this variation to better reflect soil distribution along the transect. Consider the samples near the edge of the map unit boundaries if your boundaries need to be corrected.

Exercise B

Developing a Soil Concept Map

Materials

- Hydric Soil Indicators
- Acidity, Drainage and Slope Cheat Sheet
- Map Concept Field Sheet
- Official Series Descriptions (OSDs) at NRCS web site.

Steps

1. Examine and evaluate the three support documents.
2. Review the OSDs of each Soil Series to develop an understanding of the range of characteristics represented in each of the soils.
3. Using these support documents and OSDs examine the data collected by the class along the transect and determine what soils are present and where they are located along the transect.
4. Move delineation lines as necessary and develop and diagram a toposequence concept of the soils along the transect.

8

Site Specific Input Application

In crop production mostly basic inputs are seeds or seedlings/planting material, irrigation water, fertilizers, weedicides and pesticides. Site specific input application means providing plant input requirement at right time, in right amount and at right place. Site specific is a term that refers to treating the smallest possible area as a single unit. A site is simply the smallest unit the farmer can manage with the tools available. The treatment of each site is specific to the needs of each site as determined by soil test data and crop scouting reports. There are technology tools for using site-specific management more effectively. In precision agriculture site specific application tools and technologies are as under:

Technology Tools

Global positioning system (GPS), geographic information system (GIS)-based records and data analysis, sensors, variable-rate controllers (VRC) are technology tools for nutrient management to meet crop needs and efficiently utilize available input resources.

Global Positioning System (GPS)

GPS receivers, either carry to the field or mounted on implements allow users to return to specific locations to sample or treat those areas or analysis and processing of remote sensed information.For analysis and processing the remote sensed images requires ground truth information, collected in the field, at a variety of sites and often at various times throughout the crop production season. In India use of GPS in precision farming has not picked up. In other countries GPS systems are used on planting equipment for collecting geo-referenced planting data, starter fertilizer application, and other inputs. With proper controllers, variable-rate application of inputs can be added to the management plan.

Real-Time Kinematic System (RTK)

Fixed RTK base stations, as the name suggests, have a permanent position which never changes. This position is transmitted to the tractor which can then correct its own position, which constantly varies, and align it to the base station. Therefore, creating the most accurate position. It enhances the accuracy of GPS systems by transmitting signals that correct positioning errors caused by the Earth's atmosphere. Such systems typically provide accuracy within 1 or 2 cm in position and from year to year. This enables accurate row-to-row positioning, eliminates skips and improves accuracy of harvest monitoring data.

Geographic Information System (GIS)

GIS is consisting of data and software designed for spatial analysis of GPS-referenced data. Soil survey data, soil test information, pest infestations, yield data, remote sensing imagery and other types of observations and records that can be collected and referenced with their geographic position (by GPS) are processed in GIS system. These data sets are converted to maps to illustrate their spatial variability within the field and become additional layers in the field database.

Soil Surveys

Soil surveys are done to know variabilities in soil characteristics which helps in site specific nutrient management. The Web Soil Survey provides over 2300 geographically-referenced digital soil surveys for free download from the United States Department of Agriculture (USDA) Natural Resources Conservation Service (NRCS) website. This information helps relate soil characteristics to site-specific variability observed in crop yield. USDA-NRCS Web Soil Survey provides free access to digital soil survey maps and information for all of the United States.

Intensive Soil Sampling

Precision soil sampling is one of an array of precision farming tools available to use nutrients more efficiently in agriculture. Site-specific production systems will usually require more intensive soil sampling. The most common is a 2 to 3-acre grid, preferably taken on a systematic, unaligned grid basis. It has been found that it is an advantage to shifting to a 1-acre grid, or even smaller where the field is known to be highly variable. GPS-equipped sampling systems document the precise location of the samples, so that the test results can be used to guide nutrient input decisions and to facilitate correlation with yield maps soil survey and other datasets. Now subsequent samples, fertilizer and manure applications, and crop removal can all be analysed as additional layers in the GIS database for the field, and used for calculating such details as fertilizer recommendations, nutrient use efficiency, and selected environmental parameters.

Remote Sensing

For mapping soil characteristics of the field, classification of crop species, detection of diseases, insect-pest attack, monitoring of weeds, nutrient deficiencies and mapping of crop yield remote sensed imagery can be used. Use of remote sensing in precision agriculture is influenced by the type of platforms (satellite, air or ground) used for data collection; number and width of spectral bands captured by the sensor (multi versus hyperspectral); and spatial (high, medium and low), temporal (hourly, daily and weekly) and radiometric (8-, 12- and 16-bit) resolutions at which sensors collect data. Remote sensing is becoming a useful tool for precision farming, using scanners on aircraft or satellites to monitor changes in wavelengths of light from fields and growing crops. Satellite imagery is also useful in more precise mapping of field boundaries and location of drainage lines, for example, and is often most effective when used in conjunction with field scouting (ground observations) to help identify the reasons for variability. The data collected can be mapped and analysed with the help of GIS tools, to provide additional data layers for GIS analysis and management decisions. Remote sensing helps to define the extent of problems identified in field scouting by recognizing similar patterns.

Above tools guide site specific input application in precision farming. Site-specific input application in precision farming involves the decisions on resource/ input application and agronomic practices are improved to better match soil and crop requirements as they vary in the field. Major inputs in farming are irrigation, fertilizers, weedicides and pesticides. These inputs are provided by micro-irrigation (water and fertilizers) and through precision spraying (weedicides and pesticides)

9

Irrigation and Fertigation and Nutrient Management

A. Precision irrigation

The prevalent flood irrigation in crops would be the thing of past because of limited water resources, waste of water in leaching and evaporation and cumbersome management resulting in considerable loss of water. The micro irrigation (sprinkler and drip- irrigation) are preferred depending on the field topography. Undulated fields are better irrigated by sprinkler and laser levelled field by drip irrigation. There are precise tools to deliver only required amount of water at different stages of growth of crops with the help of sensors. Irrigation water volume release is mechanised and automated. Efforts of using micro-irrigation methods such as sprinkler and drip irrigation have been made in last three decades in many parts of the world. There is no ideal irrigation method available which may be suitable for all weather conditions, soil structure and variety of crops cultures.

Precision irrigation is worldwide accepted concept in irrigation. Precision irrigation involves the accurate and precise application of water to meet the specific requirements of individual plants or management units and minimize adverse environmental impact. Commonly accepted definition of precision irrigation is sustainable management of water resources which involves application of water to the crop at the right time, right amount, right place and right manner thereby helping to manage the field variability of water in turn increasing the crop productivity and water use efficiency along with reduction in energy cost on irrigation. It utilizes a systems approach to achieve 'differential irrigation' treatment of field variation (spatial and temporal) as opposed to the 'uniform irrigation' treatment that underlies traditional management systems. Precise and regulated application of irrigation water and plant nutrients at low pressure and frequent intervals through drippers/emitters directly into the root zone of plant with the help of close network of pipes is known as drip irrigation system.

Benefits of precision irrigation

Precision irrigation has the potential to increase both the water use and economic efficiencies. Precision irrigation (drip and sprinkler) can improve application efficiency of water up to 80-90% as against 40-45% in surface irrigation.

Water savings

The primary goal of precision irrigation is to apply an optimum amount of irrigation throughout fields. The site specific or variable rate irrigation is considered as a necessary or essential component of precision irrigation. Reduction in water amount is desirable. Water savings of around 25% are possible through improvements in application efficiency obtained by spatially varied irrigation applications.

Yield and profit

Yield is better under precision irrigation in comparison to conventional irrigation owing to better water use efficiency.

B. Fertigation

Fertigation is a fertilizer application method, in which dissolved fertilizers are delivered to the crop through the irrigation system. This technology provides the opportunity to apply precise rates of water and fertilizers to the crop, and therefore, if designed correctly, can be an important precision agriculture technology.

Usually, concentrated fertilizer solutions are prepared in stock tanks, and are then injected into the irrigation water, using fertilizer injectors. The irrigation system can be drip irrigation, sprinkler system, pivot or different configurations of hydroponic systems.

Fertigation is a precision farming technology due to the following:

1. **Precise amount** – Fertigation allows precise and targeted application of water and fertilizers, to the root system.Unlike broadcast fertilizer application, drip irrigation allows for precise amounts of water and fertilizers to be applied in close proximity to the root zone of each plant.
2. **Precise timing** – Nutrients can be applied at the time they are needed by the crop.

In the absence of a fertigation system, it is difficult, or even impossible, to apply fertilizers when plants are larger and access to the field is limited. In such cases, fertilizers are often applied too early, resulting in losses of nutrients due to leaching, run-off, volatilization etc.

In most types of fertigation, this limitation does not exist. Once the fertigation system is installed, fertilizers can be easily applied at the right growing stage, according to the requirements of the crop.

The limitations for using fertigation in precision agriculture are often a result of the fertigation system design.

- The field units, or field sections are usually pre-determined. Although each of those field sections can be managed separately, it is often difficult to respond to variations within the field section.
- Except of a few start-ups, most of the existing fertigation systems do not yet have the capability of using feedback from sensors installed in the field and adjust fertilizer rates accordingly.
- When the variability between fields in the farm is high (different types of crops, different growth stages, different soil conditions etc.), having one set of stock tanks for each irrigation head is not always flexible enough in to apply precise fertilizer rates and nutrient ratios for the different field sections.

Fertigation holds a potential to be considered a precision agriculture technology. By choosing the system that is most adequate to the crop and field conditions and by designing it correctly, very high precision fertilization and irrigation can be achieved.

C. Site-Specific Nutrient Management

Site-specific nutrient management (SSNM) is a low-tech, plant need-based approach for optimally applying N, P, and K fertilizers. It enables farmers to dynamically adjust fertilizer use to fill the deficit between the nutrient needs of a high-yielding crop variety and the nutrient supply from naturally occurring indigenous sources such as soil, crop residues, manures, and irrigation water. The SSNM approach aims to apply nutrients at optimal rates and times to achieve high yields and high efficiency of nutrient use by the crop. It does not specifically aim to either reduce or increase fertilizer use. Site-specific nutrient management (SSNM) is a component of precision agriculture and can be used for any field or crop. It combines plant nutrient requirements at each growth stage and the soil's ability to supply those nutrients, and applies that information to areas within a field that require different management from the field average. Site-specific management allows for fine-tuning crop management systems along with 4R Nutrient Stewardship — the right source, rate, time and place of nutrient use. Wide spread adoption of SSNM technologies based on soil testing require extensive soil sampling and analysis which could be a hindrance considering the available infrastructure. Use of Global Positioning system (GPS) with Geographical Information System (GIS) and mapping can provide the right support as cost effective alternative.

SSNM provides guidance relevant to the context of farmers' fields. SSNM maintains or enhances crop yields, while providing savings for farmers through more efficient fertilizer use. By minimizing fertilizer overuse, greenhouse gas emissions can be reduced, in some cases up to 50%.

Advantages of SSNM

Major advantages of SSNM are listed below

1. It optimizes the supply of soil nutrients over space and time to match crop requirements.
2. SSNM increases crop productivity and improves efficiency of fertilizer use.
3. It mitigates greenhouse gases from agriculture in areas with high nitrogen fertilizer use.
4. Costly fertilizers are used only in required amount. Over and under dose is avoided.

10

Insect Pests and Disease Management

Sensing of heterogeneities in the field is the prerequisite for timely and spatially adjusted insect pests and disease management. Recent developments in sensor technologies have led to a broadening of their capacity to detect pests and thereby improve application in crop protection practice. These sensors can be space-borne (satellite), air-borne (airplane, unmanned aerial vehicle) or ground-based (handheld, vehicle-mounted) and provide spatial information that has added value management. Insects are usually unevenly distributed throughout an area in nature. Despite this knowledge, analytical methods and technology have limited insect pest management to uniform management decisions in the field. The methods and technology to map and analyse insect spatial distribution have been developed to a level where the uneven nature of insect populations can be considered for application of insect-pest management practices. Site-specific pest management utilizes spatial information about pest distribution to apply control method only where pest density is economically high within a field. This results saving in pesticide and labour cost. In conventional farming pesticides are applied uniformly to the field resulting in heavy use of pesticide and labour beside polluting the environment. Site-specific application of pesticides in precision farming helps to overcome this problem

Near-range and remote sensing techniques have demonstrated a high potential in detecting diseases and in monitoring crop stands for sub-areas with infected plants. The occurrence of plant diseases depends on specific environmental and epidemiological factors; diseases, therefore, often have a patchy distribution in the field. The use of non-invasive optical sensors for the detection, identification and quantification of plant diseases on different scales is made of in precision farming. Most promising sensor types are thermography, chlorophyll fluorescence and hyperspectral sensors. For the detection and monitoring of plant disease, imaging systems are preferable to non-imaging systems. To utilise the full potential of these highly sophisticated, innovative technologies and high dimensional, complex data for precision crop protection, a multi-disciplinary approach—including plant pathology, engineering, and informatics—is required.

Besides precision crop protection, plant phenotyping for resistance breeding or fungicide screening can be optimized by these innovative technologies. The pests and diseases groups that have to be monitored for precision crop protection differ greatly in size, patterns of incidence and epidemic spread within and between growth periods of the crops. Heterogeneity of pests and diseases often obvious and applications for site-specific control of these by targeted chemical control have reached an advanced stage. In contrast, the detection, identification and quantification of pests groups which are highly variable in space and time, is still in developmental stages. In conventional agriculture, without considering the extent of variable infestation pesticides are applied uniformly throughout the field. Which has led to many problems, besides cost. Unnecessary application of pesticides is not at all good for the environment. More over the pesticide may not be very friendly to crop growth. his necessitated the precision pest management (PPM) for different crops and their hybrids and varieties. PPM emphasizes judicious use of pesticide at mico-level where only required quantities of pesticides are applied giving due consideration of existing variabilities of pests.The use of precision in insect and plant disease management is profitable but technology intensive. It would take some time to popularise PPM keeping in view of present day agriculture and farmers in India.

11

Yield Mapping of Crops

Yield mapping or yield monitoring is a technique in agriculture of using GPS data to analyse variables such as crop yield and moisture content in a given field. This data produces a yield map that can be used to compare yield distribution within the field from year to year. This allows farmers to determine areas of the field that, for example, may need to be more heavily irrigated or are not yielding any crop at all. It also allows farmers to show the effects of a change in field-management techniques, to develop nutrient strategies for their fields, and as a record of crop yield.

Yield is ultimate indicator of variation of different agronomic parameters in different parts within the field. So mapping of yield and interpretation and correlation of that map with the spatial and temporal variability of different agronomic parameters helps in development of next season's crop management strategy. Yield monitors measure the volume or mass flow rate to generate time periodic record of quantity of harvested crop for that period. Time periodic yield data is then synchronized with location address obtained from on-board GPS system to create most common colour coded thematic map. Yield mapping can be carried out easily in mechanized crops. By using either loading cells that weighed the crop passing on a conveying belt or an array of sonic beam mounted over the grape discharge chute to estimate the volume, and the tonnage, of fruit harvested. Tree crops seem to have more stable yields. In handpicked fruits yield mapping is more difficult. Yield maps are one of the most valuable sources of spatial data for precision agriculture. In developing these maps, it is essential to remove the data points that do not accurately represent the yield at a corresponding location. Map averaging or smoothing is usually done to aid data interpretation. A long yield history is essential to avoid drawing conclusions that are affected by the weather or other unpredictable factors during a particular year. Processed yield maps can be used to investigate factors affecting the yield or to prescribe variable rate applications of agricultural inputs according to spatially variable yield goals (yield potential).

The whole process of precision farming can be divided into

- Data capturing
- Data access
- Data analysis
- Variable applications of fertilisers
- Variable applications of pesticides
- Variable seedling

Potential data sources for precision farming are:

Maps: The very important data sources for precision agriculture are field maps. These maps are typically accompanied by soil maps, parcel maps etc. Such data are digitised from existing papers maps but the data do not require density and majority of analysed parameters are not described in such maps, therefore soil samples positioned by GPS device have to be captured in the field.

GPS measurement, soil sampling and field measurement: Soil samples have to be analysed for themain nutrients to show their availability for plants insurface and in subsurface layers of soil. There are alsoother parameters which can affect the mobility ofnutrients through soil profile and soil intensity andcapacity factors. Fluctuation of individual analysedparameters is different in time and field relief.

Yield maps-measuring of crop: Yield maps are important data source for controlling the fluctuation of soil parameters. The yield maps construction is connected with spatial technological equipment on the harvesting machines.

Satellite imagery: Utilization ofremote sensing images is not easy to get by majority of farmers in villages.

Air pictures: Aerial photographs can be used more effectively than remote sensing data. Colour infrared materials are mostly used in agriculture, because plant production can be easily identified on these images.

Fertilization maps-amount of fertilization: The fertilization maps from previous season are also important information sources for evaluation of soil parameters.

Other variable applications maps: Prediction of variable application of pesticide and seedling. The whole process of precision farming is possible to divide into these parts.

Data capturing

Data access

Data analysis

Variable applications of fertilisers

Variable applications of pesticides

Variable seedling

This process can be understood as a cyclic optimisation process, with some control points. This control points are the measure of quantity and quality of the process. Different models of optimisation are used for maximum economy, minimum cost, optimisation of expenses, ecological effects etc.

12

Precision Spraying Technology

Precision farming involves the collection of data by observation and measurement, and the subsequent response. Sensing is key to the collection of data; whether this is remote from space by satellites, by aircraft or drones, or by proximal sensors used on the ground, mounted on vehicles or otherwise. Responses are increasingly automated and data are often integrated into whole-farm management systems. Early commercial applications were directed towards improving crop nutrition, however, utilising precision technologies to control weeds, pests and diseases is now a reality.

Precision crop protection requires detailed data on the occurrence and distribution of problems at the earliest possible stage of detection. An early detection is the key for successful management of pests in precision farming. Some of the key aspects are:

Identification: Identifications of crops, targets and beneficial; temporal and spatial distribution: presence of the target before, during and after the crop growth period;

Sampling techniques: statistically valid sampling regimes must be used to permit an accurate view of any problem

Thresholds: under particular circumstances, the point at which action to ensure levels of infestation or damage are kept to those deemed to be acceptable;

Prediction: the speed and pattern of development of a problem must be capable of being modelled, *e.g.* spread and severity of insect pests and fungal diseases; weed emergence according to species, dormancy and environmental conditions.

Spraying is done for the following:

A. Weed management

In agricultural crop production, indiscriminate application of herbicides leads to environmental pollution and reduced quality of the agricultural products. Variable-rate herbicide spraying (VRHS) technology can provide a solution to this problem as it can precisely adjust the application rates. It can lead towards minimal farmland pollution, safety of agricultural products and personnel, and sustainable development of agriculture.

The spraying system can be implemented mostly by adjusting the system pressure and pulse width modulation (PWM) to realize variable-rate spraying. Regarding the pressure adjustment method, when the pressure in the spray nozzle changes (high or low) it inevitably affects the atomization at the fixed nozzle tip. However, the spray flow adjustment scope is limited. Also, although effective, the PWM adjustment method can lead to discontinuous spraying under low-frequency conditions.

Precision farming helps in

a. Controlling patches of new or hard-to-control weeds to prevent their spread to the entire field.

b. Scouting fields or using remote sensing or aerial photos allows farmers to locate weed patches in the field.

c. Patches can be managed with localized spray operations, weed-sensing sprayers (in fallow land), or GPS guided systems.

d. Weed patches can also be controlled with non-chemical methods, including mowing, tillage, cutting for feeding animals.

Precision farming technologies for weed control offer the prospect of reduced herbicide use and greater profits. The profit is not large enough comparing with the complexities. However, evaluating their overall impact on profit can be very difficult. There are a number of important complexities which influence the level of benefits gained and unfortunately the complexities tend to reduce the level of benefits

The benefits weed management in precision farming are affected by a number of complexities, including the following.

(a) The benefits of the technology depend on the spatial distribution of weeds. The patchier the distribution, the greater the potential for precision weed management.

(b) The benefits of reducing herbicide usage are partly offset by the costs of increased weed competition in patches where the weed density is greater than zero, but too low to activate control by the precision weed controller;

(c) For various reasons, including inherent limitations of the technologies, the information collected by the precision weed controller will always be imperfect. Some areas of the paddock will be sprayed despite a low weed density, while some patches with many weeds will be missed.

(d) The impact of spraying or not spraying a weed is felt over subsequent years due to the level of carry-over of weed seeds.

(e) From a social perspective, the net benefits may depend on the external impacts of herbicide usage (e.g., to consumers' health, the environment, other farmers).

The basic parts of site-specific weed control technologies comprise of three key elements:

1. A weed sensing system, identifying, localising and measuring crop and weed parameters.
2. A weed management model, applying knowledge and information about crop–weed competition, population dynamics, biological efficacies of control methods and decision-making algorithms, and optimising treatments according to the density and composition of weed species, economic goals and environmental constraints.
3. A precision weed control implement, e.g. a sprayer with individual controllable boom sections or a series of controllable nozzles that enable spatially variable applications of herbicides

Some of the commercial hardware and software developed for precision agriculture practices comprise tools for weed mapping and control software that adapts spraying to local weed occurrence. However, few farmers have adopted site-specific weed management, although several studies have shown that weed occurrence and density varies significantly within a farm or a field.

B. Precision pest control

Like in weed management precision crop protection requires detailed data on the occurrence and distribution of problems at the earliest possible stage of detection. Some of the key aspects are:

- Identification: of crops, targets and beneficial.
- Temporal and spatial distribution: presence of the target before, during and after the crop growth period.
- Sampling techniques: statistically valid sampling regimes must be used to permit an accurate view of any problem.
- Thresholds: under particular circumstances, the point at which action to ensure levels of infestation or damage are kept to those deemed to be acceptable.

- Prediction: the speed and pattern of development of a problem must be capable of being modelled, e.g. spread and severity of insect pests and fungal diseases; and environmental conditions.

Knowing the Basics

The best technological developments with sprayers and applying chemicals cannot overcome poor sprayer techniques. For Count Erdal Ozkan, professor and pesticide application technology specialist at the Ohio State University, understanding your sprayer can help reduce customers' input costs and won't waste their time — or money — re-spraying the entire field. Some considerations, he says, include...

- Paying attention to spray pressure
- Setting proper boom height
- Using the correct nozzle for the specific job

Developments in Site-Specific Management & VR Sprayers

For over a decade, industry efforts to provide precise timing and placement of nutrients and chemicals without decreasing crop yields have been at the forefront. New developments in spraying technology have arrived, including

- Nozzle-by-nozzle control
- Fast-acting on-and-off nozzles
- Variable-rate tree sprayers

These developments are important to know in-depth because they help you — and your customers better understand how they can save money and time avoiding spray drifts and switching to precision sprayer technology. Top manufacturers recognized the challenges of traditional application and have begun introducing solutions, such as independent section control to auto-nozzle switching based on speed.The systems also offer A/B nozzle switching, which allows for manual or automatic switching between nozzles to better adapt to field conditions, speed, and application rates. For example, a low-drift nozzle may be used on the field edges and headlands, and then the operator could choose to use a different nozzle for spraying the interior of the field.

The value of precision application comes from five major capabilities.

1. *Application consistency at every speed*

In order to be its most effective, spray application rate must be consistent. One of the biggest challenges to maintaining this consistency comes when changing ground speed. Traditionally, sprayers adapt to changing ground speed by modifying spray pressure, however this can result in poor crop coverage. With precision application products, growers can set sprayer nozzles to auto-adjust based on the sprayer's current speed, which allows them to maintain a steady application rate.

For example, a sprayer's auto-adjustment can be set so that nozzle X applies at speeds up to 5 mph. When the sprayer goes above 5 mph, nozzle X will shut off and nozzle Y will activate. When the sprayer reaches the highest speed, for example, 12 mph, both nozzles X and Y would activate, working together to maintain the target application rate, ensuring consistency in coverage.

2. *Pressure and Droplet Size*

As with application rate, droplet size must also remain consistent for best results. Some auto-adjust systems that don't offer individual nozzle control can still adjust to the sprayer's speed, but with a change in pressure instead. At lower speeds, the system lowers the pressure, applying less product. This is only somewhat more efficient, as it also results in a change in droplet size, which then leads to inadequate coverage.

Thanks to pulse-width modulation on newer systems, desired pressure remains the same in the boom. The system instead changes the frequency and duty cycle of pulsing to adjust the rate, resulting in consistent droplet size and therefore improved coverage.

3. *Curve compensation coverage*

Precision application systems are also designed to maintain specific coverage throughout a field, even when operating along curves. This concept of "curve compensation" utilizes individual nozzle pulsing and the system's auto-adjust capabilities. When the sprayer turns, the system adjusts, automatically applying less product to the inside rows and more to the outside rows. This ensures that inside rows won't suffer from over-application and outside rows will still receive the prescribed rate. Coverage is ideal across the entire length of the boom.

4. *Skip and overlap reduction*

Individual section control is especially important when dealing with unique field characteristics such as border areas, waterways, or sensitive crops. Precision application systems provide this control by allowing individual sprayer sections to be shut off, which reduces skips and overlaps. For example, as a sprayer approaches a waterway, individual sections can be turned off, preventing product from being applied over the water. The sections then reactivate once the boom has reached an area to be sprayed, immediately pick up coverage again.

5. *Drift management*

Some crops require much finer application, or smaller droplet size; however, it can result in chemical drift to adjacent fields or crops not intended to be sprayed. This problem can be reduced with A/B nozzle switching, mentioned earlier.

When outlining a field, a grower can switch to a lower-drift nozzle, reducing the amount of material that drifts, and preventing potentially harmful effects to surrounding vegetation. When working towards the interior of the field, the grower can then switch back to a more conventional nozzle.

Additional considerations

While precision application equipment is designed to put more control in the hands of growers, it is still recommended that growers maintain a close partnership with trusted advisors, such as agronomists or dealership personnel.

Agronomists, for example, can provide information about the many products available, as well as educate growers on the various regulations that govern use, such as maximum application rate and nozzle specification. An agronomist or other knowledgeable precision product specialist can help growers follow guidelines and apply the right product at the right time.

Another good resource for growers is John Deere's GoSpray app. This app provides instructions for proper equipment setup and maintenance, giving equipment operators and technicians the opportunity to best optimize their machines. Key personnel at most dealerships are familiar with these apps and can assist growers in putting them into practice on the farm.

By offering growers greater control over product application, new precision technologies make the process more accurate, effective, and economical.

The cost of manual weed sensing was too high and it was concluded that automated weed sensing systems were a prerequisite for further progress in site-specific weed management. Another, perhaps more decisive, reason is that the technologies developed so far are dedicated to specific crops and ranges of weed species. Sensing a large, unknown number of species, while simultaneously making instantaneous decisions about the level of control, choice of herbicide etc, is still a very complex process. Dedicated systems therefore limit range of application.

UAV for spraying

In UAV spraying the crop monitoring through the multispectral camera is done which is mounted in UAV. In one flight the camera takes pictures and analyse by the geographic indicator. Based on the results it could be easy to find the area where to spray the pesticides. UAVs in precision agriculture is still in its early stage and may be a scope for further development in both the technology and the agriculture applications. Providentially, it is expended that with the development of UAV'S technology, improved image processing techniques, lower costs, flying times, batteries, new camera designs, low volume sprayers, and nozzle types. A significant number of experimental studies of UAV'S based remote sensing for agriculture application are on. Drone and crop monitoring would be inseparable in near future.

For precision agriculture, recent research has focused on the spatial variability of production environments, development of efficient and suitable data management systems, efficiency of various types of image analyses and optical sensing, efficiency of sensors and related technologies, designs of precision agriculture equipment, optimal inputs and service uses, and their spatial allocations, potentials of unmanned aerial vehicles (UAVs) and nano-technologies.

13

Weed Management

Site-specific weed control technologies practiced in precision farming are defined as machinery or equipment embedded with technologies that detect weeds growing in a crop and take action to maximise the chances of successfully controlling them taking into account predefined factors such as economics.Some of the commercial hardware and software developed for precision agriculture practices comprise tools for weed mapping and control software that adapts spraying to local weed occurrence. However, few farmers have adopted site-specific weed management, although several studies have shown that weed occurrence and density varies significantly within a farm or a field. Site-specific weed control techniques have gained interest in the precision farming community over the years. Managing weeds on a subfield level requires measuring the varying density of weeds within a field. Decision models aid in the selection and adjustment of the treatments, depending on the weed infestation. The weed control can be done either with herbicides or mechanically. A site-specific herbicide application technology can save large amounts of herbicides. Mechanical weed control techniques adapting to the weed situation in the field are applicable to a wide spectrum of crops.

A system for the discrimination of different weed species and crops from images is done which generates weed maps automatically. Models for the yield effect of weeds are developed and applied in on-farm-research experimental setups. Economic weed thresholds are derived and used for an herbicide application with a patch sprayer.Herbicide use has not eliminated weed problems in agriculture and the need for manual weed removal is still an issue within specialty crop systems, particularly during the first several weeks of crop emergence when competition for resources is greatest.

Precision agriculture is a powerful technology with potentially important applications to weed management. However, growers should remember good agronomic practices, including preventing weed infestations, managing herbicides wisely to prevent resistance, and minimizing environmental impact associated with weed management. Weeds, and methods used to control weeds, can have negative economic and environmental impacts. With precision agriculture, growers can take advantage of the patchy nature of weeds by targeting management efforts only where they are needed instead of wasting expensive and potentially hazardous inputs where weeds are not present. Weeds are patchy because weed spread, survival, and reproduction are variable within a field and over time. Weed patches stay in about the same place from year to year, even though weed density within a patch may vary

Principles of Site-Specific Weed Management. Automated site-specific herbicide application for weed control has been the focus of many recent agronomic and engineering studies because it has the potential of reducing the amount of applied chemical, thereby increasing farmer profitability and improving water quality. Site-specific herbicide application technology treats the field as a set of small management zones, where specific amounts and types of herbicide can be applied to treat the weeds present. If no weeds are present, or if the weed density is below the economic threshold for treatment (minimum weed presence that would justify the cost of treatment), no herbicide is applied.

Knowledge of the weed species present is important in order to select the most efficient herbicide. Weeds can be controlled either in real-time (tactical approach), where certain weeds have been detected, or by using a predetermined field map which identifies the species and location of weeds (strategic approach). The strategic approach requires a separate operation, but gives the grower a better estimate of the amount and type of chemical needed for a specific weed problem. Post emergence herbicide applications should be made before the weeds can cause economic yield loss. This is the critical period of weed control. This period depends on the crop, weed species, environmental conditions, and the size and density of both the crop and weed. Applying precision agriculture to weed management precision agriculture can enhance uniform herbicide application by minimizing the number of overlaps and skips and eliminating applications to non-crop areas such as waterways, wetlands, and odd-shaped boundaries. Adopting satellite-based auto-guidance (also called auto-steer) technology and using automatic boom section control can substantially reduce herbicide misapplications without requiring the operator to turn boom sections on or off. Auto-guidance allows more accurate control of the distance between two adjacent passes, significantly reducing steering-caused skips and overlaps, and does not require conventional markers. The automatic boom control shuts off individual sections of the sprayer when they are over areas that have already been treated, for example, previous pass, headland, or non-crop areas of the field. The automatic boom controller also can be programmed to spray areas that are not infested by weeds.

Site-specific herbicide management may be implemented through both soil-applied and post-emergence herbicide treatments. The rate of soil-applied herbicides (pre-plant or pre-emergence) may vary according to the spatial variation of soil texture and organic matter content. Soil texture and organic matter are relatively constant across time, so once the field is mapped for these traits, the same map can be used each year. When historic weed maps are available, it is also possible to make pre-emergence herbicide applications only to infested areas, thereby reducing herbicide costs and protecting the environment from potential chemical contamination. On the other hand, site-specific post-emergence applications can be more dynamic than pre-emergence applications and will vary in response to the weeds currently growing in the field. Historical data on weed distribution can aid decision-making, but it is beneficial to provide the controller with real-time or recent evidence of the current spatial variability of weeds. It is simple to make site-specific post-emergence burn down herbicide applications to winter or summer fallow fields using current technologies such as CropCircle®, WeedSeeker®, GreenSeeker®, or N-Sensor®1). These close proximity optical sensors use near infrared light reflectance measurements to distinguish green vegetation from bare soil and crop residue. These data can then be used to turn the sprayer on only when weeds or volunteer crops are present. Herbicide cost savings are realized from not spraying areas that lack weeds. Unfortunately, the optical reflectance characteristics of crops and many weed species are similar. The optical sensors cited here cannot be used to accurately assess spatial weed distribution after the crop is emerged. Instead, weed scouting, remote sensing (including low-altitude airborne imagery) and machine vision techniques have been used to recognize spatial weed patterns after crop emergence. Because many remote sensing and machine vision systems are not commercially practical to develop in-season weed maps, weed scouting remains the most reliable way to acquire spatial data on weed distribution. An experienced scout or operator is able to map weed infestations by species and then select herbicides to best control the weeds present.

A Practical Approach to Weed Mapping. Many data logging tools connected to a Global Positioning System (GPS) receiver can be used to record geographic coordinates of weedy areas. When a crop scout moves across a field, he or she can manually (using software interface options) mark field spots where weed presence justifies application of a certain herbicide. A geographic information system (GIS) software package can then be used to convert recorded data into a map showing target areas of a field with weed infestations requiring herbicide application. Different herbicides can be prescribed to different patches as needed. Unfortunately, most conventional tools are not easily operated when trying to record data while on the move. Various innovations (single button push, speech recognition, etc.) have been used to ease the work of a weed scout. As the scout moves across the field, he or she turns on the switches that correspond to the weed species present, and turns them off when targeted species are no longer seen. Normally, three or four weed species are mapped at a time, but extra switches allow marking field areas with either unusual species or other anomalies that may affect herbicide management. The Weed Switch Box is connected to a data logger, either directly or through a GPS receiver, and generates a digital code that records where switches are turned on and off.

Economic Return: The economic benefit of site-specific herbicide management depends on the percentage of the field infested with weeds. If more than three-fourths of the field requires a herbicide application, it is probably more economical to make a uniform application than a site-specific application because the cost of data collection to generate the weed map may equal the savings from reduced herbicide use. If less than three-fourths of the field requires a herbicide application, the cost of weed mapping and site-specific herbicide application may be less than a uniform application. However, if the herbicide cost is relatively low, environmental benefits of site-specific weed management may be more important than the economic return. In such a case, reduced environmental pollution becomes a driving adoption factor.

Essentially three kinds of technologies are in use or in engineering development for the detection, mapping, and control of weeds:

1) Visual scouting assisted by electronic recording.
2) Optical reflectance sensing systems
3) Colour and infrared machine vision with image analysis.

The success depends on the visual identification of weed species which is more dependable than machine identification. The entire process is complex involving synchronization of several technologies.

Scouting for weeds

Timely weed scouting makes it possible to know

1) what weed species are in a field
2) where they are
3) how severe the infestation is

All of these are very important to know in order to design a management program that fits the specific needs of that field.

Scout early, scout often:

Since many weed species become more challenging to manage as they get larger, scouting should start early and be repeated at several key times through the season in order to catch weeds when they are young. Herbicide applications are often not designed to be effective against mature weeds, increasing management costs.

Herbicide-resistant weeds must have herbicide application before they reach 4 inches tall in order for effective herbicides to kill the plants.

Weed species identification

Accurate weed ID is important for effective management because herbicide recommendations vary according to species, as do some mechanical, cultural, and biological strategies. Some species can look similar to each other species from afar, but may have drastically different management requirements. They should be examined closely to determine herbicide programs. Weed identification resources are available here, or by contacting your state's extension weed science specialist.

Scouting procedure

The goal is to get a representative idea of the weed populations throughout the whole field. For a 100-acre field, make 5-10 stops that are well spread out through the field. At each stop, walk 10 paces (or 30 feet) and record the following:

- Weed species present
- Life stage or height of weeds
- Lifecycle (summer annual, winter annual, perennial)
- Severity of the infestation based on number of plants (Low, medium, high)

Pre-plant scouting

Fields should be scouted prior to planting for the following reasons:

- to ensure effective burndown treatments before plants become too large to control effectively with chemicals
- to evaluate the potential need for alternative pre-plant weed management strategies such as tillage
- To ensure that crop seeds are being planted into weed-free fields

Post scouting

This should be done soon after planting to determine the efficacy of pre-treatments and to decide if further treatments are necessary. Again, particular attention should be paid to the height of the plants; weeds should not be allowed to grow past several inches tall, in order to maintain effectiveness of post treatments and minimize competition with the crop.

Scouting for resistance

Scout for weed species that are known to have herbicide resistance, especially if they have been sighted in your area or county. Failure of an herbicide application to control a weed does not necessarily mean it's herbicide-resistant. While genetic resistance must be confirmed through laboratory testing, these signs can be evidence of potential resistance:

- Spreading area of a single weed species
- Failure of an herbicide application to control a weed species that it is usually effective on at the applied dose, especially if there are other surrounding weeds are successful controlled
- Within the same weed species in the same field, there are surviving plants and controlled plants

If you suspect resistant weeds, contact your local extension agent. Collect a sample and provide them with information about the field conditions and the size of the weed population. If there are only a few plants, remove them from the field before they have a chance to spread seed.

Precision tools for weed scouting

Numerous crop scouting apps have been developed for smartphones and tablets, and some of them are particularly useful recording weed information and doing weed identification in the field.

14

Precision Agriculture Status in India and Application of Plastics in Agriculture

Precision agriculture is yet to be popularized for taking its enormous advantages. In laser land levelling and micro irrigation/ fertigation it must have made its prominence but others it is with research institutes where research is focussed on the spatial variability of production environments, development of efficient and suitable data management systems, efficiency of various types of image analyses and optical sensing, efficiency of sensors and related technologies, designs of precision agriculture equipment, optimal inputs and service uses, and their spatial allocations, potentials of unmanned aerial vehicles (UAVs), nano-technologies and several others.The adoptions of precision-agriculture technologies have, so far, mostly been limited to parts of developed countries. Major contributions to the profitability of precision agriculture is reduced costs of plant protection inputs and higher output revenues.

Application of Plastics Use of plastic is on the increase at every stage of crop production, harvest, post-harvest value addition, storage and transport to the market. Use of plastics in agriculture is termed as plasticulture. The major uses of plastic can be in the following crop production and management.

1. Water management or irrigation- water storage and distribution to crop plants water storage tanks, reservoirs, their linings, micro-irrigation sets and others.
2. Nursery production and management
3. Soil surface cover cultivation- plastic mulch
4. Crop protection nets
5. Protected cultivation- greenhouses
6. Compost production- vermi- compost bags

Part II: Protected Cultivation
(Greenhouse Technology)

15

Greenhouse Technology

Plants require proper environment for its growth and development. If environment is not suitable crop would not do well. To ensure proper environment to commercial crops for their better growth and development covered structures are used which provide microclimate appropriate for their growth and development. Growing of crops under such protective structures, also known as greenhouses, is termed as greenhouse technology. The structures harness solar radiation for photosynthesis in plants or can be provided with artificial lights (LEDs) for the purpose. The technology embraces many aspects of farming such as modelling of structures, agronomic practices for the production of the crop and the commercial viability of the venture. The technology ensures crop production in a cover more or less independent of natural prevailing environment. The covering or cladding permits desirable aspect of sun and air and help in modifying crop production environment inside. The modification of environment is aimed of creating micro-climate inside the structure/greenhouse most appropriate for the growth and development of the crop.

Hence greenhouse can be defined as a framed or inflated structure with claddings like glass polyethylene sheets, poly carbonates, fibre reinforced plastics and others in which plants are grown by creating microclimate conducive to crop for its growth and development with or without automation and mechanization. The purpose of growing crops under greenhouse conditions is to extend their cropping season and to protect them from adverse environmental conditions, such as extreme temperatures, precipitation and pests. A large number of greenhouses/poly houses have come up in almost all states of India. Though in infancy but greenhouse cultivation particularly in horticulture crops and nursery production is visible.

Principles of Greenhouse Cultivation

Greenhouse effects are the principle for greenhouse cultivation. Greenhouse is a structure made with the transparent covering material (polythene, or glass, fibre reinforced plastics, polycarbonate,) which transmit the solar energy inside the structure. This energy is absorbed by the crops and the objects inside the house releasing long wave length light which does not emit out as the cladding material is non-transparent for these light. This light gets trapped inside increasing the inside temperature. This rise in temperature in greenhouse is responsible for growing of crops in cold climate. During summer this increase in temperature can be managed by ventilation and cooling system, as in this period temperature rises beyond the critical temperature.

Greenhouse production systems in India are referred to protected or protective cultivation which covers greenhouse production having climate (mainly temperature and humidity) regulation provisions. India being energy (electricity) poor country has to depend on innovative methods to create conducive climate for plant growth and production. This is engineered in several ways and all the methods collectively called as protected or protective cultivation. Protected cultivation is defined as a cropping technique wherein microclimate surrounding the plant is modified partially or fully to suit its requirement for expression of its production potential.Protected cultivation has an important aspect of cultivation that is minimization of abiotic and biotic stress during crop production. Under protected cultivation land and water requirements are minimized to yield vegetables and other crops better and early.

Protected cultivation or greenhouse system has two major aspects of technology.

1. **Engineering Technology:** It deals with structures involving frames, cladding materials, irrigation system, tools, implements, robots, sensors and other engineering inputs which ensure optimal light, air temperature, water and plant growth requirements leading to yield higher and quality produce during on and off-season or round the year. This aspect takes care mainly of abiotic stress on the crop inside structures. In modern greenhouses computer aided mechanization and automation are key components.

2. **Improved Crop Production Technology:** It includes production package of practices with or without hi-tech involving development of high-yielding varieties and hybrids of vegetables and other crops suitable for protected cultivation conditions grown for longer period in comparison to open field cultivation. Crop harvest and post- harvest technology package is also covered under this.

Status of protected cultivation/greenhouse in India

While greenhouses have existed for more than one and a half centuries in various parts of the world, in India baring the pioneering research and development on protected cultivation of vegetables by Defence Research and

Development Organization (DRDO) during 1960s at its one of the establishments namely, Field Research Laboratory, Leh now named as Defence Institute of High Altitude Research (DIHAR), use of greenhouse technology started only during 1980's.Initially it was mainly practiced as research activity. However, in recent years in view of the globalization of international market and tremendous boost and fillip that is being given for export of agricultural produce, optimal utilization of resources like land and water and combat climate change, there has been a spurt in the demand for protected cultivation technology. The National Committee on the use of Plastics in Agriculture (NCPA, 1982) has recommended location specific trials of greenhouse technology for adoption in various regions of the country. The centre for protected cultivation technology at Indian Agricultural Research Institute, New Delhi was established in the year 1998-99 as demonstration farm and commissioned as Indo-Israel Project in January 2000 as a project undertaken jointly by the Government of India, through Department of Agricultural Research & Education (DARE) & ICAR and the Government of the State of Israel, through the Centre of International Cooperation (MASHAV) & CINADCO. The project farm was aimed to demonstrate different technologies for intensive and commercially oriented peri-urban cultivation of horticultural crops mainly flowers and vegetables for improved quality and productivity. The centre was emphasized to act as a nodal centre of research and development work and dissemination of technology on the above aspects. This centre is contributing immensely in promotion of tested technologies on protected cultivation in the country and development of human resource and imparting training. Of late there are a number of such centres in the country working in collaboration of Israel and other countries as well as independently at different state agricultural universities and private sector. Different states are promoting this greenhouse technology under Mission for Integrated Development of Horticulture (MIDH), Ministry of Agriculture and Farmers Welfare and Government sponsored programs.

In India, protected cultivation has been adopted in more than 100000 hectares till 2014-15. The states and union territories adopting protected cultivation in the country have increased from 9 states in 2007-08 to 30 states and union territories in 2014-15. The phase-wise implementation of protected cultivation has bolstered the adoption of protected cultivation across the country (http://nhm.nic.in)

Combined efforts of the national boards and schemes such as NHB (National Horticulture Board), MIDH and RKVY (*Rashtriya Krishi Vikas Yojna*) have created awareness and are providing financial assistance to the farmers to adopt protected farming methods for horticultural cultivation. The states that have consistently expanded the area under protected cultivation are

Andhra Pradesh, Gujarat, Maharashtra, Haryana, Punjab, Tamil Nadu, Himachal Pradesh and West Bengal. In India the area under the greenhouses has increased at a compound annual growth rate (CAGR) of 49.16% from 2007 to 2012. With the rising involvement of the government national boards, missions and the respective state agricultural agencies, the area under protected cultivation must have risen at a CAGR of 84.2% by 2017. The subsidy schemes, international associations with countries such as Israel are expected to act as a major boost for the protected cultivation in India. The only major challenge for the adoption of the protected cultivation is the cost of the structures and lack of awareness and production knowledge. If these aspects are taken care of by the scientists, industrial players and the policy makers, then the protected cultivation industry will showcase remarkable growth in near future (www.marketresearch.com) as it has done in China.

Table: Effect of poly house on productivity of vegetable crops as compared to open cultivation

	Open cultivation Mt/ha			Poly house cultivation Mt/ha			
Crop	Min.	Max.	Average	Min.	Max.	Average	% increase
Capsicum	25	30	27.5	90	150	120	336
Tomato	40	45	42.5	180	250	215	406
Cucumber	15	20	17.5	80	100	90	414
Beans	10	15	12.5	24	32	28	124
Peas	10	15	12.5	20	25	22.5	80
Coriander	10	12	11.0	15	20	17.5	59
Spinach	10	15	12.5	20	25	22.5	80

(Department of Soil Science and Water Management, Dr Y S Parmar, University of Horticulture and Forestry, Nauni, Solan 173 230, India)

Advantages of protected cultivation/greenhouse system:

There are several advantages of adopting protected cultivation of vegetables, flowers, plant nurseries in different agro climatic regions of the country. Some of the major advantages of protected cultivation are as under:

1. There is large increase in yield, produce quality and revenue under protected cultivation of horticultural crops.

2. Since, generally, drip irrigation and/ or fertigation practices are followed under large scale crop production under protected cultivation, water evaporation rate is reduced and weed population is minimal, high water use efficiency is there. There is saving in significant amounts of water under protected cultivation. Advantage of water harvesting from roof tops of structure is also there.

3. Use of pesticides, generally, is less under protected cultivation and there is significant reduction in pesticide use leading to reduction in production costs which results in healthier produce.

4. Off- season and year-round production of particular crops is possible under protected cultivation practices. This allows farmers to take advantage of market seasonality and off season higher prices. It is well established advantage of protected cultivation.

5. Protected cultivation is an attraction to educated unemployed youths in villages and pre-urban areas which is helpful in attraction of them to farming.

6. Affordable low-cost poly-houses or required structures such as poly tunnel, rain shelter and others are fabricated locally by small-scale entrepreneurs leading to local employment generation.

7. Hi-tech greenhouse cultivation methods such as soilless cultivation, hydroponics, vertical farming, aeroponics, aquaponics, hi-tech plant nursery production in soilless medium, balcony gardens and others are pre-urban and unban farming of tomorrow which are to be promoted through protected cultivation in the country. These technologies are known as new age protected cultivation technologies of crops particularly for horticultural crops such as vegetables, flowers, medicinal plants, mushrooms and new group of edibles microgreens.

Protected cultivation adds pride to the farming profession which is otherwise low in India.

16

Types of Greenhouses

Different types of greenhouses or protected structures are designed to meet the specific needs. They can be classified based on shape, utility, construction, cladding material, net houses and technology adopted.

A.Classification based on shape of greenhouse: Greenhouses can be classified based on their shape or style. For the purpose of classification, the uniqueness of the cross section of the greenhouses can be considered as a factor. As the longitudinal section tend to be approximately the same for all types, the longitudinal section of the greenhouse cannot be used for classification. The cross sections depict the width and height of the structure and the length is perpendicular to the plane of cross section. Also, the cross section provides information on the overall shape of the structural members, such as truss or hoop, which will be repeated on every day. The commonly followed types of greenhouse based on shape are lean-to, even span, uneven span, ridge and furrow, saw tooth and quonset.

Lean-to type greenhouse: A lean-to design is used when a greenhouse is placed against the side of an existing building. It is built against a building, using the existing structure for one or more of its sides. It is usually attached to a house, but may be attached to other buildings. The roof of the building is extended with appropriate greenhouse covering material and the area is properly enclosed. It is typically facing south side. The lean-to type greenhouse is limited to single or double-row plant benches with a total width of 7 to 12 feet. It can be as long as the building it is attached to. It should face the best direction for adequate sun exposure. The advantage of the lean-to type greenhouse is that, it usually is close to available electricity, water, and heat. It is a least expensive structure. This design makes the best use of sunlight and minimizes the requirement of roof supports. It has the following disadvantages:

limited space, limited light, limited ventilation and temperature control.

The height of the supporting wall limits the potential size of the design. Temperature control is more difficult because the wall that the greenhouse is built on, may collect the sun's heat while the translucent cover of the greenhouse may lose heat rapidly. It is a half greenhouse, split along the peak of the roof.

Even span type greenhouse: The even-span is the standard type and full-size structure, the two roof slopes are of equal pitch and width. This design is used for the greenhouse of small size, and it is constructed on level ground. It is attached to a house at one gable end. It can accommodate 2 or 3 rows of plant benches. The cost of an even-span greenhouse is more than the cost of a lean-to type, but it has greater flexibility in design and provides for more plants. Because of its size and greater amount of exposed glass area, the even-span will cost more to heat. The design has a better shape than a lean-to type for air circulation to maintain uniform temperatures during the winter heating season. A separate heating system is necessary unless the structure is very close to a heated building. It will house 2 side benches, 2 walks, and a wide centre bench. Several single and multiple span types are available for use in various regions of India. For single span type the span in general, varies from 5 to 9 m, whereas the length is around 24 m. The height varies from 2.5 to 4.3 m.

Uneven span type greenhouse: This type of greenhouse is constructed on hilly terrain. The roofs are of unequal width; make the structure adaptable to the side slopes of hill. This type of greenhouses is seldom used now-a-days as it is not adaptable for automation.

Ridge and furrow type greenhouse: Designs of this type use two or more A-frame greenhouses connected to one another along the length of the eave. The eave serves as furrow or gutter to carry rain and melted snow away. The side wall is eliminated between the greenhouses, which results in a structure with a single large interior, Consolidation of interior space reduces labour, lowers the cost of automation, improves personal management and reduces fuel consumption as there is less exposed wall area through which heat escapes. The snow loads must be taken into the frame specifications of these greenhouses since the snow cannot slide off the roofs as in case of individual free standing greenhouses, but melts away.

Saw tooth type Greenhouse: These are also similar to ridge and furrow type greenhouses except that, there is provision for natural ventilation in this type. Specific natural ventilation flow path develops in a saw- tooth type greenhouse.

Quonset greenhouse: This is a greenhouse, where the pipe arches or trusses are supported by pipe purling running along the length of the greenhouse. In general, the covering material used for this type of greenhouses is polyethylene. Such greenhouses are typically less expensive than the gutter connected greenhouses and are useful when a small isolated cultural area is required. These houses are connected either in free, standing style or arranged in an interlocking ridge and furrow. In the interlocking type, truss members overlap sufficiently to allow a bed of plants to grow between the overlapping portions of adjacent houses. A single large cultural space thus exists for a set of houses in this type, an arrangement that is better adapted to the automation and movement of labour.

Gothic Arch Greenhouse

Gothic arch greenhouses are similar to the Quonset. They are made up of two separate curved pieces that meet at the ridge line of the roof. Their walls show less pronounced curvature. As a result, compared with the Quonset, these offer better space and headroom along the walls and the design lends itself to more efficient ventilation. The Gothic arch style is preferred for its aesthetic appeal.

Greenhouses are available in various shapes and sizes as listed with drawings below.

(i)Spherical dome, (ii) Hyperbolic paraboloid, (iii) Quonset, (iv) Modified quonset, (v)Gothic arch, (vi) Mansard roof, (vii) Even span, and (viii) uneven span. The even span greenhouse is most common in application.

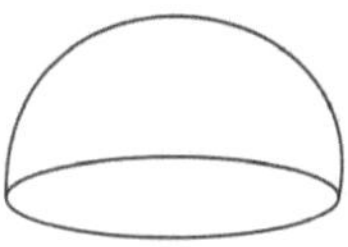

(i) Spherical dome

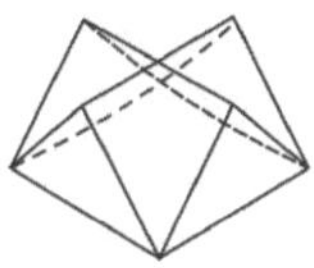

(ii) Hyperbolic paraboloid

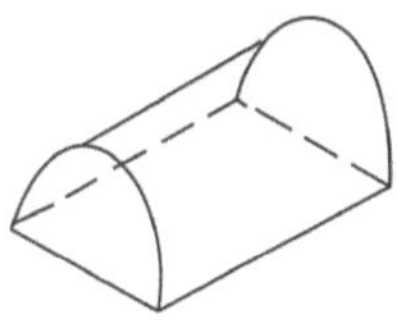

(iii) Quonset

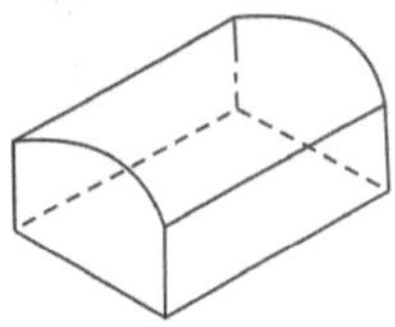

(iv) Modified quonset

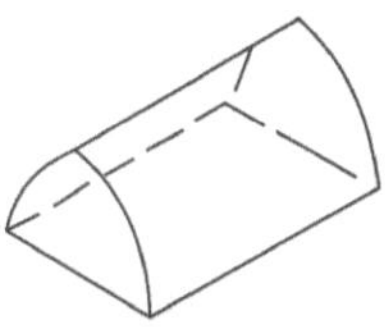

(v) Gothic arch

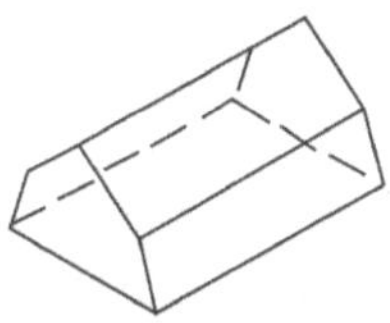

(vi) Mansard roof

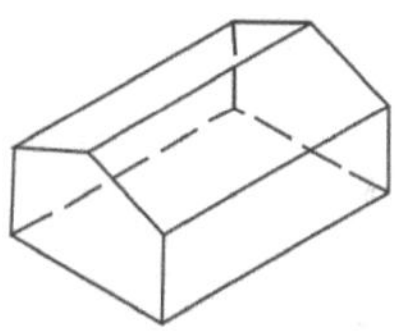

(vii) Even span

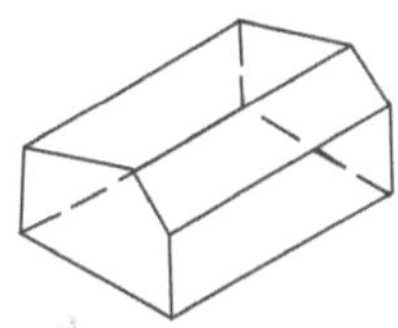

(viii) Uneven span

B. Classification based on utility of greenhouse

Classification of greenhouses can be made depending on the functions or utilities. Of the different utilities, artificial cooling and heating of the greenhouse are more expensive and elaborate. Hence based on the artificial cooling and heating, greenhouses are classified as green houses for active heating and active cooling system.

Greenhouses for active heating: During the night time, air temperature inside greenhouse decreases. To avoid the cold bite to plants due to freezing, some amount of heat has to be supplied. The requirements for heating greenhouse depend on the rate at which the heat is lost to the outside environment. Various methods are adopted to reduce the heat losses, viz., using double layer polyethylene, thermo pane glasses (two layers of factory sealed glass with dead air space) or to use heating systems, such as unit heaters, central heat, radiant heat and solar heating system.

Greenhouses for active cooling: During summer season, it is desirable to reduce the temperatures of greenhouse than the ambient temperatures, for effective crop growth. Hence suitable modifications are made in the greenhouse so that large volumes of cooled air are drawn into greenhouse, this type of greenhouse either consists of evaporative cooling pad with fan or fog cooling. This greenhouse is designed in such a way that it permits a roof opening of 40% and in some cases nearly 100%.

On the basis of working and principal applications, greenhouse may be broadly classified in the three categories: (i) greenhouse for crop cultivation, (ii) greenhouse for crop drying and (iii) greenhouse as the solar energy collector. The detailed classification on this basis is shown in

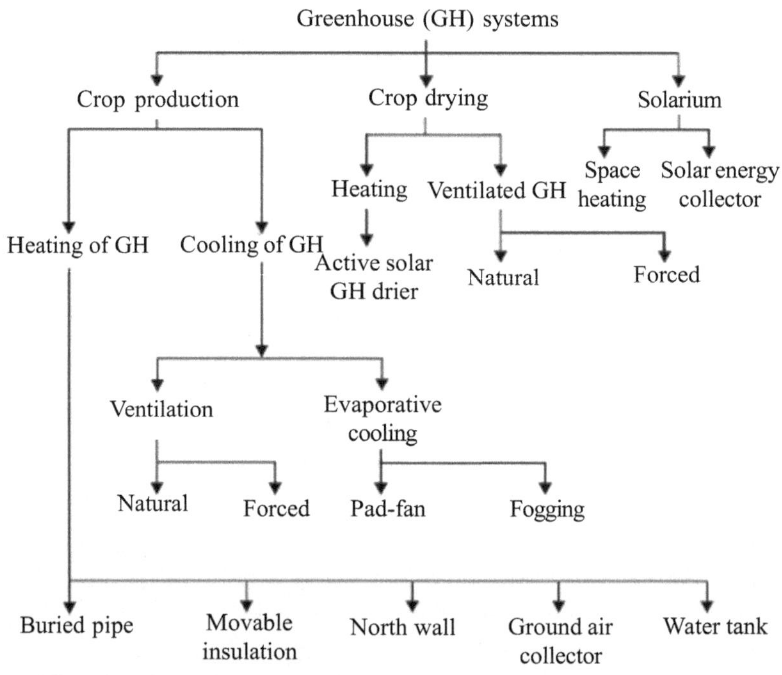

C. Classification based on framing materials

The classification may also be based on the material of the construction of the frame. Aluminium, galvanized steel, wood, reinforced concrete, and bamboo are the commonly used framing materials; aluminium being the most economical. Aluminium framing is longer lasting, relatively maintenance free, and allows for light reflectance from the frame. Steel is commonly used but must be galvanized to resist high moisture conditions within the greenhouse. Steel needs more maintenance than aluminium and is heavier. Wood has been the traditional framing material but it is not readily available, is expensive, and requires considerable preservative treatments against natural and pest-based decaying. Bamboo is gaining popularity because it is cultivated, environment friendly, low cost, reasonably strong, and amenable to mechanical fabrication and maintenance. Bamboo based greenhouse frames have been developed at different institutes in the country as low cost greenhouse.

The foundation of a greenhouse column consists of a concrete pole cast in a concrete block at a certain depth below the surface. The dimensions of these concrete blocks depend on a number of parameters, such as the greenhouse height, the number of connected spans, the number of stability bracings, the soil type, etc.

The type of construction is predominantly influenced by the structural material, though the covering material also influences the type. Span of the house indicates the selection of structural members and their construction. Higher the span, stronger should be the material and more structural members are used to make sturdy truss type frames. For smaller spans, simpler designs like hoops can be followed. Therefore, based on construction, greenhouses can be broadly classified as wooden framed, pipe framed and truss framed structures.

Wooden framed structures: In general, for the greenhouses with span less than 6 m, only wooden framed structures are used. Side posts and columns are constructed of wood without the use of a truss. Pine wood and bamboo are commonly used as they are inexpensive and possesses the required strength. Locally available timber, with good strength, durability and machinability also can be used for the construction.

Pipe framed structures: Pipes are used for construction of greenhouses, when the clear span is around 12m. In general, the side posts, columns, cross ties and purlins are constructed using pipes. In this type, the trusses are not used.

Truss framed structures: If the greenhouse span is greater than or equal to 15m, truss frames are used. Flat steel, tubular steel or angular iron is welded together to form a truss encompassing rafters, chords and struts. Struts are support members under compression and chords are support members under tension. Angle iron purlins running throughout the length of greenhouse are bolted to each truss. Columns are used only in very wide truss frame houses of 21.3 m or more. Most of the glass houses are of truss frame type, as these frames are best suited for pre-fabrication.

D.Classification based on covering/cladding materials

Most of the greenhouses were constructed using glass as the glazing material/ claddings prior to World War II so that the term glasshouse and greenhouse were synonymous. With the development of plastic films and rigid sheets began the era of plastic greenhouses. Polyethylene, PVC, EVA, acrylic, polycarbonate, fiberglass, and PVF are some of the plastic materials which have been used as greenhouse cladding materials. A greenhouse glazing material is expected to meet the specifications related to the following parameters for a given greenhouse.

- Transmission for solar radiation (UV and PAR)
- Transmission for heat radiation (NIR and FIR)
- Thermal insulation value
- Resistance to aging due to UV radiation
- Moisture permeability
- Mechanical strength (tensile and impact)
- Fire resistance
- Costs, and
- Available dimensions

Glass, rigid plastic sheets, and plastic films are the three categories of greenhouse glazing materials. Glass is considered the best material from radiation properties point of view. Glass has the highest light transmission and could last up to 30 years or more. However, glass is heavy, brittle, and expensive. Strength and safety considerations suggest that tempered or laminated glass should be used. Polycarbonate, fibre-glass, and acrylic are the common rigid plastics used for greenhouse glazing. Rigid plastics last from 7 to 20 years and are less expensive than glass. Polycarbonate and acrylic are usually manufactured as twin-walled sheets. Air in the space between the two walls acts as an insulator. Light transmission through rigid plastics is very good initially, but decreases over time as the plastics age and turn yellow due to UV component in sunlight. The rigid plastic sheets are much lighter than glass, requiring lighter greenhouse frames.

Plastic films (e.g., PE, PVC, EVA) are the least cost options of greenhouse glazing among the whole range of greenhouse glazing materials, but they usually last only three to four years. Plastic films for greenhouses are normally 150 to 200 micron thick. When used in two layers with air-inflation under cold conditions, there is approximately 60 % reduction in heating requirement as compared to that for a single layer glass or plastic house. A common additive to the film material (IR absorber) helps in reducing the heat loss from the greenhouse during cold ambient conditions. Some films are manufactured with a special surface treatment to prevent condensation droplets from falling on the crop (so-called no-drip films). The radiation transmission characteristics of different glazing materials have been summarized in Table.

Table: Transmittance for direct and diffuse light of greenhouse glazeing materials (PAR)

Material	Thickness	Transmittance for direct light	Transmittance for diffuse light
Glass	4 mm	89-91%	82%
PE film	200 μm	89-91%	81%
EVA	180 μm	90-91%	82%
PVC	200 μm	87-91%	79%
ETFE	100 μm	93-95%	88%
PC	12 mm	80%	61%
PMMA	16 mm	89%	76%
PC zigzag	25 mm	88%	79%

Clearly, ETFE (Ethylene tetrafluoroethylene), EVA, and PC zigzag have PAR properties close to or better than glass. The greenhouse warming effect, however, needs FIR transmission reduced for the glazing materials. Comparison of greenhouse glazing materials on the basis of transmittance for long wave heat radiation is given in Table.

Table: Transmittance for long wave heat radiation (FIR)

Material	Thickness	Transmittance
Glass	4 mm	0
PE	200 μm	40-60
PE-IR	200 μm	20-40
EVA	180μm	20-40
ETFE	100 μm	15-20
PC	12 mm	0
PMMA	16 mm	0
Knitted polyester white	140 g m^{-2}	22
Knitted polyester white	150 g m^{-2}	11
Knitted polyester black		5
Polyester textile white		12
Polyester textile trans		approx 50
Polyester textile Alum		3
Black		0

Plastic film greenhouses have some disadvantages. Generally, the effort required for installing and replacing the film may be excessive. The film quality deteriorates due to ageing caused by UV radiation and friction against the structure. Film, if not anchored properly, flutters against the structure and, thus, weakens faster over time. Poor condensation behaviour of the film may cause falling of water drops from the film onto the plant leaves causing pathogenic conditions. Ventilation is poor in multi-span film greenhouses. Dust deposition on plastic glazing is a serious problem; it could reduce solar energy transmission considerably.

Covering materials have direct influence on the greenhouse effect inside the structure and they alter the air temperature inside the house. The types of frames and method of fixing also varies with the covering material. Based on the type of covering materials, the greenhouses are classified as glass, plastic film and rigid panel greenhouses.

Glass greenhouses: Only glass greenhouses with glass as the covering material existed prior to 1950. Glass as covering material has the advantage of greater interior light intensity. These greenhouses have higher air infiltration rate which leads to lower interior humidity and better disease prevention. Lean-to type, even span, ridge and furrow type of designs are used for construction of glass greenhouse.

Plastic film greenhouses: Flexible plastic films including polyethylene, polyester and polyvinyl chloride are used as covering material in this type of greenhouses. Plastics as covering material for greenhouses have become popular, as they are cheap and the cost of heating is less when compared to glass greenhouses. The main disadvantage with plastic films is its short life. For example, the best quality ultraviolet (UV) stabilized film can last for four years only. Quonset design as well as gutter-connected design is suitable for using this covering material.

Rigid panel greenhouses: Polyvinyl chloride rigid panels, fibre glass-reinforced plastic, acrylic and polycarbonate rigid panels are employed as the covering material in the Quonset type frames or ridge and furrow type frame. This material is more resistant to breakage and the light intensity is uniform throughout the greenhouse when compared to glass or plastic. High grade panels have long life even up to 20 years. The main disadvantage is that these panels tend to collect dust as well as to harbour algae, which results in darkening of the panels and subsequent reduction in the light transmission. There is significant danger of fire hazard.

Shading nets: There are a great number of types and varieties of plants that grow naturally in the most diverse climate conditions that have been transferred by modern agriculture from their natural habitats to controlled crop conditions. Therefore, conditions similar to the natural ones must be created for each type and variety of plant. Each type of cultivated plant must be given the specific type of shade required for the diverse phases of its development. The shading nets fulfil the task of giving appropriate micro-climate conditions to the plants.

Shade nettings are designed to protect the crops and plants from UV radiation, but they also provide protection from climate conditions, such as temperature variation, intensive rain and winds. Better growth conditions can be achieved for the crop due to the controlled micro-climate conditions "created" in the covered area, with shade netting, which results in higher crop yields. All nettings are UV stabilized to fulfil expected lifetime at the area of exposure. They are characterized of high tear resistance, low weight for easy and quick installation with a 30-90% shade value range. A wide range of shading nets are available in the market which are defined on the basis of the percentage of shade they deliver to the plant growing under them.

Insect-proof net/ screen houses: Insect-proof screens are widely used in agriculture for passive microclimate control and for insect exclusion. It is an efficient tool for crop production in adverse climatic and environmental condition. As screens are made of a porous material, the protected environment usually interacts with the outside and hence screens provide only moderate microclimatic modifications. Such modifications might be crucial for certain horticultural crops, thus may strongly influence production and quality. Depending on the type of screen, structure configuration, crop and climatic region, it has been shown that compared to open field conditions, screens reduce solar radiation and air velocity by about 15–39% and 50–87%, respectively; increase air relative humidity by 2–21%; decrease air temperature and evapotranspiration by 2.3–2.5 °C and 17.4–50% respectively. Besides protecting crops from insects-pests' microclimate helps in improved productivity.

E. Classification based on technology used

Greenhouses are a technology based investment. The higher the level of technology used, the greater potential for achieving tightly controlled growing conditions. This capacity to tightly control the conditions in which the crop is grown is strongly related to the health and productivity of the crop. The following three categories of greenhouse have been defined to assist people in selecting the most appropriate investment for their needs and budget.

1. Low technology greenhouses

These greenhouses are less than 3 metres in total height. Tunnel houses, or "igloos", are the most common type. They do not have vertical walls. They have poor ventilation. This type of structure is relatively inexpensive and easy to erect. Little or no automation is used.

While this sort of structure provides basic advantages over field production, crop potential is still limited by the growing environment and crop management is relatively difficult. Low level greenhouses generally result in a suboptimal growing environment which restricts yields and does little to reduce the incidence of pests and diseases. Pest and disease control, as a result, is normally structured around a chemical spray program.

Low technology greenhouses have significant production and environmental limitations, but they offer a cost effective entry to the farming.

2. Medium technology greenhouses

Medium level greenhouses are typically characterised by vertical walls more than 2m but less than 4 metres tall and a total height usually less than 5.5 metres. They may have roof or side wall ventilation or both. Medium level greenhouses are usually clad with either single or double skin plastic film or glass and use varying degrees of automation.

Medium level greenhouses offer a compromise between cost and productivity and represent a reasonable economic and environmental basis for the industry. Production in medium level greenhouses can be more efficient than field production. Hydroponic systems increase the efficiency of water use. There is greater opportunity to use non-chemical pest and disease management strategies but overall the full potential of greenhouse horticulture is difficult to attain.

3. High technology greenhouses

High level greenhouses have a wall height of at least 4 metres, with the roof peak being up to 8 metres above ground level. These structures offer superior crop and environmental performance. High technology structures will have roof ventilation and may also have side wall vents. Cladding may be plastic film (single or double), polycarbonate sheeting or glass. Environmental controls are almost always automated. These structures offer enormous opportunities for economic and environmental sustainability. Use of pesticides can be significantly reduced. High technology structures provide a generally impressive sight and, internationally, are increasingly being involved in agribusiness opportunities.

Although these greenhouses are capital intensive, they offer a highly productive, environmentally sustainable opportunity for an advanced fresh produce industry. Investment decisions should, wherever possible, look to install high technology greenhouses. These greenhouses are suitable for modern technologies such as hydroponics, aeroponics, vertical farming and microgreen production.

17

Typical Applications of Greenhouses

Application of greenhouses in agriculture is mainly to modify microclimate for the production of crops with multiple objectives of raising crops in season offseason, round the year, early crops, late crops, breed crops, raise genetically engineered plants, seed production so and so forth. With hi-tech agriculture and modern techniques greenhouses are used for growing crops anywhere and everywhere, on soil under soil, water, under water and space. Greenhouses are mainly used for crop nursery production, crop production largely in vegetables, flowers and herbs. Among fruits mainly strawberry is grown under greehouses commercially. Other fruits can also be grown.

Besides crop production greenhouses are used for dehydration or drying of produce, seed production, for crop improvement using biotechnology and other sciences. Use of greenhouses in uncongenial environments for crop production is common in places like high altitudes, deserts, arctic climate and alike. One of the largest greenhouse complexes in the world is in Almería, Andalucía, Spain, where greenhouses cover almost 200 km2 (49,000 acres). One of the important use of greenhouses world over is raising tissue culture plants through a popularly known process hardening. For hardening of tissue culture plants different types of low and high technology greenhouses are must. Greenhouses in hot, dry climates used specifically to provide shade are sometimes called shade houses. Hydroponic crop production is generally done in greenhouses. Greenhouses are also used as retail market centres mainly for ornamental plants and seedlings and grafts of horticultural crops. Greenhouses are used for maintaining bee (honey bees and bumble bees) colonies as crop pollinators.

As explained above the major applications of greenhouses are listed below:

Major Applications of Greenhouses

Modern greenhouses are generally used for the following:

a) In plant nursery multiplication and rooting of planting materials /cuttings of horticultural crops/plants/grafting of horticulture plants.

b) In forcing or growing greenhouse crops mainly horticulture crops.

c) In ordinary crops greenhouses for the growing of vegetables, herbs, flowers and fruits.

d) In drying greenhouses for the drying process of agricultural products and forestry products.

e) In hydroponic greenhouses for soilless crop production – aeroponics and aquaponics.

f) Vertical farming

g) Microgreens production

h) Urban horticulture/food factories/ indoor farming

i) Hydroponic fodder production

j) Breeding new varieties of crops and raising genome edited plants

k) Vertical horticulture- mushrooms, orchids and others

l) Seed production of several horticultural crops

m) Tissue culture plants hardening and mass seedling production of these.

n) Butter fly gardens

o) Research on plant protection and physiology

p) Conservation of certain germplasm

To start greenhouse farming high investment is required. Hence though the crop is having high commercial value and sustainable market demand but for want of resource this crop is cultivated in open field with several risks. Floriculture crops like all cut flower, foliage plants, vegetables and herbs are mostly grown in the greenhouse where return is high due to off-season cultivation or high yield or healthy crops or several other factors. The profitability of greenhouse crop depends upon the various factors such as market, climatic condition, labour, raw material availability.

The following crops are commercially grown in greenhouses in India. Vegetables are also the most common crops grown with protected agriculture followed by ornamental crops.

Vegetables: tomato, cherry tomato, cucumber, coloured capsicum, lettuce and number of other leafy Indian and exotic vegetables and button mushrooms.

Floriculture: gerbera, Dutch roses, carnations, *Anthurium*, lily, Orchids, Gypsophila, Gladiolus, Chrysanthemum, number of foliage and flowering plants.

Fruits: Strawberry, papaya and other small berries

Herbs: basil, chives, oregano, cilantro, parsley and several others.

Plant nursery production: seedlings of vegetables and flowers- hybrids and grafts, saplings of fruit crops and forest tree species, sugarcane and paddy nurseries and nurseries of other several foliage and ornamental crops.

Microgreens: young vegetable greens that are approximately 1–3 inches (2.5–7.5 cm) tall.

Greenhouses are playing major role in food supply of high altitude countries like in Almería, Andalucía, Spain, where greenhouses cover almost 200 km^2 (49,000 acres). Even in Iceland they play major role in food supply and greenhouse heating source there is geothermal. In fact, several fruit crops like mango, citrus fruits, guava etc are being grown in climate controlled greenhouses or under several protected structures.

18

Choice of Crops for Cultivation Under Greenhouses

Many crops can be grown in greenhouse but the choice depends on number of factors.

1. Requirement of the farmer, what for she/he wants to grow the crop.
2. For personal use number of vegetable crops and herbs and fruits like strawberry and papaya can be grown.
3. In case of commercial crops, the market must be ascertained before the choice of the crop.
4. Greenhouse must be equipped to create micro-climate for the crop of choice.
5. The choice of crop is a crucial factor for the economic feasibility and success of a commercial greenhouse operation. The revenue of cultivated crop makes the venture to progress.
6. Crop varieties, suitable for greenhouse cultivation should differentiate themselves from open-field varieties to fetch better price on quality.

Based on the experience in India and abroad the following crops can be the choice in greenhouse cultivation for commercial purpose. Most of the vegetable crops, herbs and floriculture crops and their nurseries can be grown in one or the other type of greenhouse. Greenhouses are ideally suited for raising seedlings or nurseries even of paddy, wheat and sugarcane etc. The following plants thrive better in a greenhouse environment, and are most profitable crops if market is assured.

Tomatoes and cherry tomatoes: The most common greenhouse crop, tomatoes can also be quite profitable. Many mostly indeterminate varieties and hybrids do well under different greenhouses

Lettuce: The most common and successful and large grown crop in greenhouses. Varieties with different colour and shape do well in low and high tech. green houses. Liking of this crop in India is limited to reputed hotels and very high income group people.

Peppers: Peppers are also known as Capsicums or Shimla mirch and are quite remunerative crop of greenhouses in different colour- green, red, yellow and purple varieties.

Cucumbers: Seedless or parthenocarpic hybrids or varieties of cucumbers do very well under greenhouses and are popular in India also both in off and on season.

Leafy greens: In addition to lettuce large number of leafy vegetables do well in off and on season in different greenhouses.

Green beans: Grown in greenhouses, are in high demand, especially in the inner city, selling for incredible prices.

Squash: Different varieties of squash as an early crop is beneficial in most of the greenhouses. Pollination requirement is to be met.

Microgreens: Microgreens are in demand because of their better nutritive value.Around 7 days old seedlings of different crops mainly of mustard family are now extensively grown under greenhouse.

Herbs: Some of the most popular herbs such as basil, mint, chives, oregano, cilantro and parsley, and others do well in greenhouses.

Strawberries, raspberry and papaya: A lucrative commercial crop for greenhouses different methods of cultivation such as flat beds, soilless, hydroponics etc. Certain fruit crops with low canopy natural or managed planted in high density do well under different types of greenhouses, Shade net and insect screens are becoming popular in fruit crop farming.

Ornamental crops: Besides nursery plants and crop seedlings a large number of floriculture crops do well under greenhouses and grown keeping in view the local demand for better price. The most common crops grown under greenhouses are cut flowers such as rose, lilies, gerbera, carnation, anthurium, gypsophila, orchids, Limonium, Alestroemeria, gardenias, snapdragons, asters, pansies, annual chrysanthemums, dahlias and many more.

Fodder: Several fodder crops (oat, maize and others) for green fodder are hydroponically grown under greenhouses.

Spices: Several spices being high value crops can be grown under greenhouses. Greenhouse production of spices is gradually picking-up as very lucrative venture.

Monoculture (growing only crop) or two crops or more than two crops in a same greenhouse can be grown keeping in view their temperature, humidity and other environmental requirements.

Selection of crop should be done keeping in view the type of greenhouse, the facilities available and the local market demand of the crop.

19

Plant Response to Greenhouse Environment

The productivity of a crop is result of genetic make-up (heredity) and environment (climate) and their interaction. The components of crop environment or climate or microclimate are light, temperature, air compositions and the nature growing medium or soil. In open fields, only manipulation of nature of the soil or root medium by tillage, irrigation and fertilizer application is possible. Under greenhouse structure control of any one or more of the components of the micro climate is manipulated.

Light

The visible light of the solar radiation is a source of energy for plants. Light energy, carbon dioxide (Co2) and water all enter in to the process of photosynthesis through which carbohydrates are formed. The production of carbohydrates from carbon dioxide and water in the presence of chlorophyll, using light energy is responsible for plant growth and reproduction. The rate of photosynthesis is governed by available fertilizer elements, water, carbon dioxide, light and temperature. The photosynthesis reaction can be represented as follows

$$CO_2 + \text{water} + \text{light energy} \xrightarrow[\text{Plant nutrients}]{\text{Chlorophyll}} \text{carbohydrates} + \text{oxygen}$$

Considerable energy is required to reduce the carbon that is combined with oxygen in CO_2 gas to the state in which it exists in the carbohydrate. The light energy thus utilized is trapped in the carbohydrate. If the light intensity is diminished, photosynthesis slows down and hence the growth. If higher than optimal light intensities are provided, growth again slows down because of the injury to the chloroplasts. The light intensity is measured by the international unit known as Lux. It is direct illumination on the surrounding surface that is one meter from a uniform point source of 1 international candle. Green house crops are subjected to light intensities varying from 129.6klux on clear summer days to 3.2 Klux on cloudy winter days. For most crops, neither condition is ideal. Many crops become light saturated, in other words, photosynthesis does not increase at light intensities higher than 32.2klux. Rose and carnation plants will grow well under summer light intensities. In general, for most other crops foliage is deeper green if the greenhouse is shaded to the extent of about 40% from mid spring (May) to mid fall (August and September).

Thus, it is apparent that light intensity requirements of photosynthesis vary considerably from crop to crop. Light is classified according to its wave length in nanometres (nm). Not all light useful in photosynthesis process. UV light is available in the shorter wavelength range, i.e less than 400nm. Large of quantities of it is harmful to the plants. Glass screens are opaque to the most UV light and light below the range of 325nm. Visible and white light has wavelength of 400 to 700nm. Far red light (700 to 750nm) affects plants, besides causing photosynthesis. Infrared rays of longer wavelengths are not involved in the plant process. It is primarily, the visible spectrum of light that is used in photosynthesis. In the blue and red bands, the photosynthesis activity is higher, when the blue light (shorter wavelength) alone is supplied to plants, the growth is retarded, and the plant becomes hard and dark in colour. When the plants are grown under red light (longer wavelength), growth is soft and internodes are long, resulting in tall plants. Visible light of all wavelengths is readily utilized in photosynthesis

Temperature

Temperature is a measure of level of the heat present. All crops have temperature range in which they can grow well. Below this range, the plant life process stop due to ice formation within the tissue and cells are possibly punctured by ice crystals. At the upper extreme, enzymes become inactive, and again process essential for life cease. Enzymes are biological reaction catalyst and are heat sensitive. All biochemical reactions in the plant are controlled by the enzymes. The rate of reactions controlled by the enzyme often double or triple for each rise of temperature by 10 degrees Celsius, until optimum temperature is reached. Further, increase in temperature begins to suppress the reaction and finally stop it. As a general rule, green house crops are grown at a day temperature, which are 3 to 6 degrees Celsius higher than the night temperature on cloudy days and 8 degrees Celsius higher on clear days. The night temperature of greenhouse crops is generally in the range of 7 to 21 degrees Celsius. *Primula*, *Mathiola incana* and *Calceolaria* grow best at 7 degrees Celsius, carnation and cineraria at 10 degrees Celsius, rose at 16 degrees Celsius, chrysanthemum and poinsettia at 17 to 18 degrees Celsius and African violet at 21 to 22 degrees Celsius.

Relative humidity

As the green house is a closed space, the relative humidity of the greenhouse air will be more when compared to the ambient air, due to the moisture added by the evapo-transpiration process. Some of this moisture is taken away by the air leaving from the greenhouse due to ventilation. Sensible heat inputs also lower the relative humidity of the air to some extent. In order to maintain the desirable relative humidity levels in the greenhouses, processes like humidification or dehumidification are carried out. For most crops, the acceptable range of relative humidity is between 50 to 80 per cent. However, for plant propagation work, relative humidity up to 90 per cent may be desirable. In summer, due to sensible heat addition in the daytime, and in winters for increasing the night time temperatures of the greenhouse air, more sensible heat is added causing a reduction in the relative humidity of the air. For this purpose, evaporative cooling pads and fogging system of humidification are employed. When the relative humidity is on the higher side, ventilators, chemical dehumidifiers and cooling coils are used for dehumidification.

Ventilation

A green house is ventilated for either reducing the temperature of the greenhouse air or for replenishing carbon dioxide supply or for moderating the relative humidity of the air. Air temperatures above 35 degrees Celsius are generally not suited for the crops in green house. It is quite possible to bring the greenhouse air temperature below this upper limit during spring and autumn seasons simply by providing adequate ventilation to the green house. The ventilation in a greenhouse can either be natural or forced. In case of small greenhouses (less than 6m wide) natural ventilation can be quite effective during spring and autumn seasons. However, fan ventilation is essential to have precise control over the air temperature, humidity and carbon dioxide levels.

Carbon dioxide

Carbon is an essential plant nutrient and is present in the plant in greater quantity than any other nutrient. About 40 per cent of the dry matter of the plant is composed of carbon. Under normal conditions, carbon dioxide (CO_2) exits as a gas in the atmosphere slightly above 0.03 per cent or 345ppm. During the day, when photosynthesis occurs under natural light, the plants in a greenhouse draw down the level of CO_2 to below 200ppm. Under these circumstances, infiltration or ventilation increases carbon dioxide levels, when the outside air is brought in, to maintain the ambient levels of CO_2. If the level of CO_2 is less than ambient levels, CO_2 may retard the plant growth. In cold climates, maintaining ambient levels of CO_2 by providing ventilation may be uneconomical, due to the necessity of heating the incoming air in order to maintain proper growing temperatures. In such regions, enrichment of the green house with CO_2 is followed. The exact CO_2 level needed for a given crop will vary, since it must be correlated with other variables in greenhouse production such as light, temperature, nutrient levels, cultivar and degree of maturity. Most crops will respond favourably to CO_2 at 1000 to 1200 ppm.

Greenhouse environment consisting of mainly the above mentioned parameters in optimal range of the requirement of a crop would govern the performance and profit.

20

Growing Medium

Soil is the main growing medium. Natural as well as synthetic substitutes to soils are cropping up. Under greenhouse cultivation growing medium is generally top soil as used in open field/traditional farming. The practices of soil preparation for planting crops are more or less same as under open field. Generally, crops are planted on raise beds. As under open conventional farming soil preparation are made as per the crops or their varieties/hybrids to be grown. To enrich the soil with nutrients soil is tested for its physical and chemical properties and amended accordingly keeping in view the crop requirement. The soil composition is checked with each crop cycle and amended as per requirement of next crop. To avoid the harmful microbes and insects build up in the soil it is advisable to solarize or fumigate the greenhouse soil each year during hottest month of the year. Soil solarisation is an environment friendly soil pathogen management method and practiced widely in north India.

Soil Solarisation

Soil solarisation is an environment friendly method of using solar thermal power for controlling pests such as soil borne plant pathogens including fungi, bacteria, nematodes, and insect and mite pests along with weed seeds and seedlings in the moist soil by mulching the soil usually with a transparent polyethylene cover, to trap solar energy in greenhouses. It may also be described as a method of decontaminating soil or soil disinfestation using sunlight or solar thermal power. This energy/temperature causes desirable physical, chemical, and biological changes in the soil.

Fumigation in greenhouse

The accessories in greenhouse such as the propagation room, containers, flats, knives, working surface, benches etc. can be disinfected using one part of formalin in fifty parts of water or one part sodium hypochlorite in nine parts of water. An insecticide such as dichlorvos and others recommended from time to time sprayed regularly will take care of the insects present if any. Care should be taken to disinfect the seed or the planting materials before they are moved into the greenhouse with a recommended seed treatment chemical for seeds and a fungicide –insecticide combination for cuttings and plugs respectively. Disinfectant solution such as tri-sodium phosphate or potassium permanganate placed at the entry of the greenhouse would help to get rid of the pathogens from the personnel entering the greenhouses. It is advisable to use botanicals where ever possible. In hi-tech greenhouse it is advisable to have disinfect facility at entry point of workers, supervisors and managers. Use of permissible fungicides would be safer option.

Soil Mixes

Soil mixes used for greenhouse production of potted plants and cut flowers are highly modified mixtures of soil, organic and inorganic nutrients. When top soil is included as a portion of the mixture, it is generally combined with other materials to improve the water holding capacity and aeration of the potting soil. Many greenhouses do not use topsoil as an additive to the soil mixes, but rather use a combination of these organic and inorganic components as an artificial soil mix. When managed properly as to watering and fertilization practices, these artificial mixes grow crops that are equal if not better to those grown in top soil. The media used in greenhouse generally have physical and chemical properties which are distinct from field soils.

Characteristics of growing medium

- A desirable medium should be a good balance between physical properties like water holding capacity and porosity.
- The medium should be well drained.
- Medium which is too compact creates problems of drainage and aeration which will lead to poor root growth and may harbour disease causing organisms.
- Highly porous medium will have low water and nutrient holding capacity, affects the plant growth and development.
- The media reaction (pH of 5.0 to 7.0 and the soluble salt (EC) level of 0.4 to 1.4 dS/m is optimum for most of the greenhouse crops.
- A low media pH (<5.0) leads to toxicity of micronutrients such as iron, zinc, manganese and copper and deficiency of major and secondary nutrients while a high pH (>7.5) causes deficiency of micronutrients including boron.
- A low pH of the growth media can be raised to a desired level by using amendments like lime (calcium carbonate) and dolomite (Ca-Mg carbonate) and basic, fertilizers like calcium nitrate, sodium nitrate and potassium nitrate.

- A high pH of the media can be reduced by amendments like sulphur, gypsum and Epsom salts, acidic fertilizers like urea, ammonium sulphate, ammonium nitrate, mono ammonium phosphate and aqua ammonia and acids like phosphoric and sulphuric acids.
- It is essential to maintain a temperature of the plug mix between 21 to 26 degree Celsius. Irrigation through mist is a must in plug growing. Misting for 12 seconds every 12 minutes on cloudy days and 12 seconds every 6 minutes on sunny days is desirable.
- The pH of water and mix should be monitored regularly.

Desirable nutrient level in greenhouse growth media

S.No.	Category	Concentration (mg/l)			
		NO3	N	P	K
1.	Transplants	75	125	10-15	250-300
2.	Young pot & foliage plants	50	90	6-10	150-200
3.	Plants in beds	125	225	10-15	200-300

Media ingredients and mix

Commercially available materials like peat, sphagnum moss, vermiculite, perlite and locally available materials like sand, red soil, common manure/ compost, coco peat and rice husk can be used in different proportions to grow greenhouse crops. These ingredients should be of high quality to prepare a good mix. They should be free from undesirable toxic elements like nickel, chromium, cadmium, lead etc.

Pasteurization of greenhouse plant growing media

Greenhouse growing medium may contain harmful disease causing organisms, nematodes, insects and weed seeds, so it should be decontaminated by heat treatment or by treating with volatile chemicals like methyl bromide, chloropicrin etc.

Agent	Method	Recommendation
Heat	Steam	30 min at 180° F
Methyl bromide	10 ml/cu. ft. of medium	Cover with gas proof cover for 24-48 hr. Aerate for 24-28 hr before use.
Chloropicrin	(Tear gas) 3-5 ml/cu. ft. of medium	Cover for 1-3 days with gas proof cover after sprinkling with water. Aerate for 14 days or until no odour is detected before using.
Basamid	8.0 g/cu.ft. of medium	Cover for 7 days with gas proof cover and aerate for atleast a week before use.
Formalin	20 ml/l of water (37%)	Apply 2 l/cu.ft. cover for 14 to 36 hr and aerate for at least 14 days.

Disinfection of the growing media can also be achieved by fungicides or bactericides

Fungicides and their effect on a few fungi

Chemical	Rate of application	Effect against
Captan	2 g/l of water	*Pythium, Fusarium, Rhizoctonia and Phytophthora. Some extent to root and stem rot, white mold,black rot, crown rot and damping off.*
Metalaxyl + Mancozeb (Ridomil MZ 72 WP)	1 g/l of water	*Pythium, Phytophthora, Fusarium* and other soil borne pathogens

Temperature necessary to kill soil pests

- 115°F for water molds (*Pythium* and *Phytophthora*)
- 120°F for nematodes
- 135°F for worms, slugs and centipedes
- 140°F for most plant pathogenic bacteria
- 160°F for soil insects
- 180°F for most of weed seeds
- 200°F for few resistant weed seeds and plant viruses

Growing media

1) **Soil** – Heavy, requires sterilizaation, high CEC
2) **Sand** – No CEC, inert, heavy used for improving aeration
3) **Peat moss** – Sphagnum or hypnum peatmoss (~70% decomposed)
4) **Vermiculite** – Expanded mica mineral high water holding, good CEC, good buffering
5) **Perlite** – Expanded volcanic silicaceous rock (1400° F, vacuum) very light, inert, no CEC, no buffering
6) **Pumice** – Crushed volcanic rock (heavier than perlite) similar properties as perlite
7) **Bark, Wood shavings and sawdust** – Problem with phenolics, tannins. Use after decomposition
8) **Rockwool** – Shredded, sheets, cubes, no CEC, no buffering, inert.

Soil Mixes for Container growing

Heavy soil: Soil, perlite or sand and peat in 1:2:2 ratio

Medium soil (silt loam): Soil, Serlite or sand and peat in 1:1:1 ratio

Light Soil (sandy loam): Soil and peat in 1:1 ratio

Fertilizers added to above keeping in view the requirement of plants to be raised.

Before use growing media is to be sterilized using steam or chemicals.

It is advisable to check pH of medium before use and adjust using sulphur powder, aluminium sulphate, iron sulphate, and acids like sulphuric acid, phosphoric acid, nitric acid to lower the pH. Use lime ($CaCO_3$), or Dolomite ($CaCO_3 \cdot MgCO_3$) to raise the pH

21

Planning and Design of Greenhouses

Planning

A greenhouse, is basically the purpose of providing and maintaining a growing environment that will result in optimum production at maximum yield. The agriculture in the controlled environment is possible in all the regions irrespective of climate and weather. It is an enclosing structure for growing plants, greenhouse must admit the visible light portion of solar radiation for the plant photosynthesis and, therefore, must be transparent. At the same time, to protect the plants, a greenhouse must be ventilated or cooled during the day because of the heat load from the radiation. The structure must also be heated or insulated during cold nights. A greenhouse acts as a barrier between the plant production areas and the external or the general environment.

Site selection and orientation

A greenhouse is designed to withstand local wind, snow and crop loads for a specific cropping activity. In this way, the structure becomes location and crop specific. The building site should be as level as possible to reduce the cost of grading, and the site should be well aerated and should receive good solar radiation. Provision of a drainage system is always possible. It is also advisable to select a site with a natural windbreak. In regions where snow is expected, trees should be 30.5 m away in order to keep drifts back from the greenhouses. To prevent shadows on the crop, trees located on the east, south, or west sides should be at a distance of 2.5 times their height.

Site Selection for greenhouse

For selecting site for greenhouse following are considered and paid serious attention.

1. **Solar radiation:** Since plants require sunlight for photosynthesis areas or location with high light intensity year-round are desirable. With more of cloudy days' crop yield in tomato, cucumber and capsicum is likely to be adversely affected.

2. **Water**: Main input in crop production is water. Its quality and quantity both are important. For irrigation in conventional way would require about 5 litre per plant per day in crop like tomato. For evaporative cooling also huge quantity of water is necessary. For ferti-gation (fertilizer and watering) also water requirement is there. To minimize water requirement drip irrigation coupled with plastic mulch be used wherever feasible.
3. **Elevation:** Heating and cooling cost, if required, are affected with elevations.
4. **Wind:** Heavy wind areas, besides high cost of construction is risky to wind storm and other related conditions
5. **Tree:** Trees should not be nearer to the structure to avoid shade and danger of damaging the structure with falling leaves and branches following strong wind.
6. **Snow:** In high hills heavy snowing areas be avoided.
7. **Dust:** Greenhouse should not be adjusting to national high ways or dusty roads.
8. **Pests prevalence:** As far as possible avoid having structure in areas infested with aphids, spider mites, thrips and white fly. These are major pests in greenhouses.
9. **Stable and level ground:** Greenhouse should be built at compact and levelled ground with slight inclination say 4 inch drop in lave at100 feet.
10. **Utilities:** Day to day utilities availability be ensured in the vicinity including easy access by road for ferrying inputs and produce.
11. **Orientation:** East-west orientation ensures better light during winter.In multispan greenhouse gutter be oriented North-South. In naturally ventilated poly houses ventilator should open leeward side. A free standing greenhouse should have its long axis perpendicular to the wind direction.
12. **Expansion:** Always have a provision of future date expansion of greenhouse on ground.
13. **Availability of labour and inputs:** Ensure availability of adequate labourer and inputs required for farming in the locality. Great distance for procuring these.
14. **Others:** Create ancillary services and structures such as office, stores, pack house etc.

Structural design

Greenhouse structures are designed by detailed consideration of the loads on the individual components and connections through a rational engineering analysis, either by hand or computer. Greenhouse structures need to be analysed for all building code required load conditions. Loads are commonly classified as dead loads, live loads, snow loads, wind loads and in some locations, earthquake loads. The standards for greenhouse construction in different countries are given in the following Table.

Table: Greenhouse Design Standards in Different Countries

Country	Standards	Remarks
USA	NGMA Standards for Design Loads in Greenhouse Structures	National Greenhouse Manufacturers Association (NGMA) is a voluntary association of prominent U.S. greenhouse manufacturers (NGMA, 2010)
European Union	EN13031-1Greenhouse design and Construction Part I: Commercialproduction greenhouses	Based on structural Eurocodes, ENV1991-ENV1999 andcomplementary information (Elsner et al, 2000)
France	NFU 57060, NFU 57063, NFU 57064	Roux and Gratraud. 1990
India	IS 14462	Voluntary standard developed by the Bureau of Indian Standards
Japan	Standard for the structures of greenhouses	Japan Greenhouse Horticulture Association, Tokyo (Moriyama, 2015)
South Korea	Korean Building Code-Structural Korean Greenhouse Design Standards	Ministry of Land, Infrastructure, and Transport of Korea
Germany	DIN 11535-2	(Elsner et al, 2000)
Australia	AS/NZS 1170	Jointly published by Standards Australia, GPO Box 5420, Sydney, NSW 2001 andStandards New Zealand, Private Bag 2439, Wellington 6020
The Netherlands	NEN 3850	Calculation principles of buildings wind and snow loads (NEN, 1996)

Role of Tradition

Tradition also plays an important role in the development of greenhouse facility at a given location. Successful designs of the past usually remain in use for a long time even though improved products become available in the market. The lack of compatibility between old and new products is usually the main reason restricting the use of newer technologies. This resistance to the adoption of improved technologies may be termed ad technology-inertia. Unless there are overriding considerations, the users continue with older technologies. Generally, a greenhouse grower begins the production by using simple technology. After the initial success, the greenhouse grower invests the profits in building better-equipped, more sophisticated greenhouses. However, a level of compatibility between the modern and the old greenhouses is maintained.

An understanding of different types of loads in the context of greenhouse structure is presented in the following sections.

Dead loads

Dead loads are gravity loads that are constant in magnitude throughout the life of the building. They include the walls, roof, glazing and fixed equipment. Fans, suspended heaters and overhead piping for water or heat are also considered a dead load, as they are permanently attached to the structure. Long-term crops, such as tomatoes or cucumbers that are suspended from the trusses, are frequently identified as a dead load. Any permanent loads such as hanging baskets, planters, etc., that are to be supported by structural members for an extended time period of more than 30 days will also be included as part of the dead load. Short-term crops, such as hanging baskets, are normally considered a live load.

Live loads

Live loads are more difficult to calculate, as they can change. One common live load is a maintenance crew repairing glazing on the roof. Another is a trolley conveyor that moves plants into the greenhouse. The carrier full of plants weighs 90 kg to 140 kg. This is a concentrated load that moves as the conveyor is pushed along. The National Greenhouse Manufacturers Association (NGMA) design guidelines recommend that purlins, rafters and trusses be designed to support a minimum concentrated load of 45 kg at mid-span. Short-term plant loads are considered live loads. NGMA recommends that these be limited to a maximum of 750 Pa. A crop of 25 cm hanging baskets could add from 150 Pa to 250 Pa. Any live load if it is imposed on the structure for a continuous period of 30 days or more shall be considered permanent and, therefore, included as part of the dead load. Pitched and arched greenhouse roofs shall be designed to safely support the minimum live load.

All roof members such as purlins, rafters, truss top members, etc., are required to safely support a minimum concentrated live load of 45 kg applied downward and normal to the roof surface at their mid span. The bottom chord panel points of roof trusses should safely support a minimum concentrated live load of 45 kg applied at any panel point. The concentrated live load includes adequate allowance for ordinary impact conditions such as monkey jumping.

Greenhouse manufacturer needs to inform the greenhouse owner of the live loads for which the greenhouse has been designed and the greenhouse owner must ensure that a live load greater than that for which the roof or roof supporting members were designed is not placed upon the roof or supporting members.

Wind loads

Loading from the wind can come from any direction, but is usually considered to act in a horizontal direction against the walls. Basic wind speeds for design are 160 kmph. The basic wind speed is adjusted for factors such as site exposure; height and shape of the building; roof slope; and use factor. Greenhouses and sales buildings that are open to the public have a higher use factor than a production greenhouse. The above factors are applied to the wind velocity to get a wind load. Structural member size and resistance to overturning are then calculated.

Local structural elements in a greenhouse which are directly loaded by the wind are glass, rigid plastics, or fiberglass glazing materials and the connection devices used to attach these materials to the structure. Secondary members that support the glazing materials and transfer the wind loads to main wind-force resisting system (members such as purlins and lintel beams) should be considered as components. Design wind loads for greenhouses consider basic wind speed, velocity pressure, and the design wind pressure. Values for wind uplift must not exceed the dead load of the structure; otherwise the structure could be lifted out of the ground. This can occur occasionally to high tunnels that only have the pipes driven 45-60 cm into the soil. These structures weigh very little, and the hoop shape acts like an airplane wing that tends to create uplift when the wind passes over it.

Snow loads

Snow loads vary considerably from 0 Pa to more than 4.79 k Pa depending upon the location. Local building codes specify the design snow load. Snow can be light and fluffy with a water equivalent of 30 cm equal to 2.5 cm of rain. When wet and heavy, 30 cm of snow may equal 7.5 to 10 cm of rain. Snow having a 2.5 cm rainwater equivalent will load a structure with 250 Pa.

Design snow loads are usually adjusted to consider building exposure, roof slope, heat loss through the roof and the type of occupancy. As the heat loss through the glazing on a heated greenhouse is high, the design usually considers that most of the snow will melt or slide off. Drifting snow causes imbalanced loads that may lead to greenhouse collapse. Two to four times normal loads may occur where drifts or sliding snow phenomena are observed. It can occur on the leeward bays of a gutter-connected range. A space of about 3 m should be left between freestanding greenhouses to provide space for the snow that slides off.

Seismic Load

International Building Code exempts production greenhouses from seismic design requirements. However, earthquake loads imposed on the greenhouse structure by non-structural components, such as, mechanical equipment, irrigation equipment, etc. need to be considered. Lateral force-resisting systems are needed to meet seismic detailing requirements and limitations. All parts of the structure should be interconnected. These connections are designed to resist the seismic force induced by the parts being connected. Any smaller portion of the structure should be tied to the remainder of the structure with a connection capable of transmitting an appropriate value of load to a larger portion of the structure.

Other Loads

For the sake of completeness, although not very relevant, structural designs consider flood and hydrostatic stresses. Stresses might arise due to thermal expansion / contraction affecting frame joints and / or glazing material. There are situations where rainwater may create additional stresses. For example, in a Quonset greenhouse, if the glazing film is loosely applied, water may begin ponding at some locations during rains, causing the failure of the glazing film or the structure.

Generally, combinations of loads are used for greenhouse design as several types of loading may occur at the same time. For example, there may be a high wind associated with a snowstorm. In some areas, earthquake loads may have to be considered. However, Production greenhouses are usually exempt, as they have limited human occupancy. Greenhouse manufacturers need to provide a structural analysis for the intended greenhouses. Computer-aided design software allows faster, easier and more reliable designs. There are powerful computer software that perform greenhouse designs reliably and precisely.

The most important function of the greenhouse structure and its covering is the protection of the crop against hostile weather conditions (low and high temperatures, snow, hail, rain and wind), diseases and pests. It is important to develop greenhouses with a maximum intensity of natural light inside. The structural parts that can cast shadows in the greenhouse should be minimized. The different structural designs of greenhouse based on the types of frames are available. A straight side wall and an arched roof is possibly the most common shape for a greenhouse, but the gable roof is also widely used. Both structures can be free standing or gutter connected with the arch roof greenhouse. The arch roof and hoop style greenhouses are most often constructed of galvanized iron pipe. If tall growing crops are to be grown in a greenhouse or when benches are used, it is best to use a straight side wall structure rather than a hoop style house, this ensures the best operational use of the greenhouse. A hoop type greenhouse is suitable for low growing crops, such as lettuce, or for nursery stock which are housed throughout the winter in greenhouses located in extremely cold regions. A gothic arch frame structure can be designed to provide adequate side wall height without loss of strength to the structure. Loads in designing the greenhouse structures include the weight of the structure itself and, if supported by the structure, loads of the equipment/fan/heaters for the heating and ventilation and water lines. Greenhouse structures should be designed to resist a 130 km/h wind velocity. The actual load depends on wind angle, greenhouse shape and size, and the presence or absence of openings and wind breaks.

Utilities

Any greenhouse facility would need electricity and water. As indicated earlier, even if the greenhouses are naturally ventilated, electricity would be needed in head house, sales area and storage section. Water is required for irrigation, general conveniences, and environmental control. The requirements of water and electricity are being considered in this section.

Water

Adequate supply of good quality water is necessary. The amount of water requirement depends upon the area under production, crop, and climate. A general rule is to make a provision for about 20 L / m^2 for watering daily if growing in ground beds. In summers, watering could be done more frequently. The provision of water be higher if evaporative cooling method is going to be used. Water is also needed to meet the requirements of people working there. Not only the available water is judiciously used, efforts must be made to harvest whatever rain water is available.

Electricity

A naturally ventilated greenhouse does not need any power supply. But, in general, for greenhouse electrical appliances, such as motor fans, lights, and other equipment we need assured electrical power for uninterrupted operations. Therefore, adequate provision on this account should be made. An auxiliary generating system should be available to cope up with the failure of regular electrical supply. Table 1 gives the requirement of power for different sizes of greenhouses.

Table: Electrical power requirements for different types of greenhouses

Greenhouse size (m^2)	Electrical power requirement (amp/volts)
500 - 1000	60/240
1000 - 2000	100/240
2000 - 3000	150/240
3000 - 4000	200/240
4000 - 8000	400/240
8000 - 12000	600/240

Profitability of greenhouse operation could be increased by minimizing the input costs through efficient resource use. Solar power should be harnessed to the maximum extent. Crop residues and other effluents need to be value-added for their reuse.

The most important function of the greenhouse structure and its covering is the protection of the crop against hostile weather conditions (low and high temperatures, snow, hail, rain and wind), diseases and pests. It is important to develop greenhouses with a maximum intensity of natural light inside. The structural parts that can cast shadows in the greenhouse should be minimized. The different structural designs of greenhouse based on the types of frames are available. A straight side wall and an arched roof is possibly the most common shape for a greenhouse, but the gable roof is also widely used. Both structures can be free standing or gutter connected with the arch roof greenhouse. The arch roof and hoop style greenhouses are most often constructed of galvanized iron pipe. If tall growing crops are to be grown in a greenhouse or when benches are used, it is best to use a straight side wall structure rather than a hoop style house, this ensures the best operational use of the greenhouse. A hoop type greenhouse is suitable for low growing crops, such as lettuce, or for nursery stock which are housed throughout the winter in greenhouses located in extremely cold regions. A gothic arch frame structure can be designed to provide adequate side wall height without loss of strength to the structure. Loads in designing the greenhouse structures include the weight of the structure itself and, if supported by the structure, loads of the equipment/fan/heaters for the heating and ventilation and water lines. Greenhouse structures should be designed to resist a 130 km/h wind velocity. The actual load depends on wind angle, greenhouse shape and size, and the presence or absence of openings and wind breaks.

The ultimate design of a greenhouse depends on the following aspects:

1. The overall structural design and the properties of the individual structural components.

2. The specific mechanical and physical properties which determine the structural behaviour of the covering materials.
3. The specific sensitivity of the crop to light and temperature to be grown in the greenhouse.
4. The specific requirements relevant to the physical properties of the covering material.
5. The agronomic requirements of the crop.

Covering/cladding materials

The following factors are to be considered while selecting the greenhouse covering material i.e., light, transmission, weight, resistant to impact, and durability to outdoor weathering and thermal stability over wide range of temperatures. Before selecting the covering material, two important points should be taken into consideration: the purpose for which greenhouse facility is intended and service life of material. In temperate regions where high temperatures are required, the covering material with high light transmission and far IR absorption must be selected. Also the loss of heat by conduction should be minimum.

Covering material life span

1. Glass and acrylic sheet 20 years
2. Polycarbonate and fiberglass-reinforced polyester sheet 5-12 years
3. Polyethylene/plastic 2-6 months
4. Polyethylene stabilized for UV rays 2-3 years

The ideal greenhouse selective claddings should have the following properties:

- It should transmit the visible light portion of the solar radiation which is utilized by plants for photosynthesis.
- It should absorb the small amount of UV in the radiation and convert a portion of it to fluoresce into visible light, useful for plants.
- It should reflect or absorb IR radiation which are not useful to plants and which causes greenhouse interiors to overheat.
- Should be of minimum cost and should have usable life of 10 to 20 years

Bureau of Indian Standards (BIS) has recently revised specifications on cladding materials **(IS 15827 : 2019).**

22

Design Criteria of Greenhouse for Cooling and Heating Purposes

Greenhouses should provide a controlled environment for plant production with sufficient sunlight, temperature and humidity. Greenhouses need exposure to maximum light, particularly in the morning hours. Consider the location of existing trees and buildings when choosing your greenhouse site. Water, fuel and electricity make environmental controls possible that are essential for favourable results. For this reason, use reliable heating, cooling and ventilation.

The greenhouse temperature requirements depend upon which plants are to be grown. Most plants require day temperatures of 70 to 80 degrees F, with night temperatures somewhat lower. Relative humidity may also require some control, depending on the plants cultured. Some plants grow best in cool greenhouses with night temperatures of 50 degrees F after they are transplanted from the seeding tray. These plants include azalea, daisy, carnation, aster, beet, calendula, camellia, carrot, cineraria, cyclamen, cymbidium orchid, lettuce, pansy, parsley, primrose, radish, snapdragon, sweet pea and many bedding plants. Some plants grow best in warm greenhouses with night temperatures of 65 degrees F. These plants include rose, tomato, poinsettia, lily, hyacinth, cattleya orchid, gloxinia, geranium, gardenia, daffodil, chrysanthemum, coleus, Christmas cactus, calla, caladium, begonia, African violet, amaryllis and tulip. Tropical plants usually grow best in high humidity with night temperatures of 70 degrees F.

General definitions

Air circulation: The process of moving or mixing air within a greenhouse to control temperature, humidity and carbon dioxide distribution.

Aspirate: To circulate air continuously across or through an object, such as a temperature sensor.

Cooling: Generally, the removal of heat from the interior of the greenhouse; however, the term may also be applied to the conversion from sensible to latent energy, as in evaporative cooling.

Evaporative cooling: The addition of moisture to air (usually ventilation air) to reduce its dry bulb temperature. Sensible energy is converted to latent energy with negligible change in the enthalpy of the air.

Glazing: The transparent or translucent covering of a greenhouse; e.g., glass, plastic film, rigid plastic panels, and others

Heating: The addition of heat to the interior of the greenhouse from any energy source, including the sun.

Horizontal air circulation: A system utilizing fans to generate a horizontal air circulation pattern above and through the plant canopy.

Infiltration: Uncontrolled air exchange which occurs through small, uncontrolled openings in the greenhouse covering, such as gaps around doorways, utility service entrances and between cover sections. These

exchanges are driven by wind pressure and/or temperature differentials inside and outside the greenhouse. Infiltration rate is generally expressed in terms of internal air volume changes per unit of time (e.g., air exchanges per hour).

Fan ventilation: Air exchange through side wall and/or roof openings, which occurs when fans are used to move air into, and exhaust air out of, the greenhouse. Fans may be located either at the inlet end (positive pressure) or the exhaust end (negative pressure); however, the most common location is at the exhaust end.

Natural ventilation: Air exchange which occurs in response to temperature and pressure variations inside and outside the green house. These variations are created and maintained by solar energy, internal heat sources, and/or wind.

Ridge and furrow: A type of greenhouse construction where modular units are connected at the gutters to cover large ground areas; also called gutter-connected greenhouses.

Ventilation: The process of exchanging air inside the greenhouse with outside air to control temperature, humidity and carbon dioxide levels.

Ventilation rate: The volume of air exchanged per unit of time per unit floor area. Ventilation rate is often expressed as m3/(s·m2) (cfm/ft2) of greenhouse floor area.

Greenhouse Cooling

There are various active and passive ways of cooling the greenhouses which are briefly explained below:

Ventilation

Air exchange between greenhouse environment and outside environment is known as ventilation. This exchange maintains inside temperature of greenhouse nearer to outside temperature. There are two systems of ventilation in greenhouses.

1. Natural ventilation

2. Mechanical ventilation

1. *Natural Ventilation:* the process of supplying and removing air through an indoor space without using mechanical systems. It refers to the flow of external air to greenhouse as a result of pressure or temperature differences. The benefits of natural ventilation are

1. Improved air quality in greenhouse
2. Reduction of greenhouse gas emissions
3. Increased worker's efficiency

Unfortunately, natural ventilation does not work as efficiently on warm days when the outside wind velocity is low (less than 200 feet per minute). Keep in mind that whether using either a mechanical (fans only) or naturally ventilated system with no other cooling capabilities (i.e., without evaporative cooling pads or a fog system), the indoor temperature cannot be lowered below the outdoor temperature.

2. *Mechanical Ventilation:* Circulate air using ducts and fans rather than relying on airflow through limited natural spaces in greenhouse. The benefits of mechanical ventilation are

1. Better indoor air quality
2. Provides proper fresh air flow along with appropriate locations for intake and exhaust
3. Allows constant flow of outside air with filtration.

Some greenhouses can be ventilated using side and ridge vents, which run the full length of the greenhouse and can be opened as needed to provide the desired temperature. This method uses thermal gradients, creating circulation due to warm air rising. Greenhouses with only side vents depend upon wind pressure for ventilation and are usually not satisfactory. The warm air must be allowed to rise through the ridge vent while cooler air enters along the sides. The vent size is important. Ridge vents should be about one-fourth the floor area and the side vents about the same size. The roof vents should open above the horizontal position to provide about a 60-degree angle to the roof. Most of these vents are manually operated.

Greenhouse design considerations for cooling

Due to the long and narrow design of most freestanding greenhouses, mechanical ventilation systems usually move the air along the length of the greenhouse (the exhaust fans and inlet openings are installed in opposite end walls). Natural ventilation systems provide crosswise ventilation (using side wall and roof vents). In gutter-connected greenhouses, mechanical ventilation system inlets and outlets can be installed in the side or end walls. Natural ventilation systems usually consist of only roof vents. Ultimate natural ventilation systems include open-roof greenhouse designs, where the very large ventilation openings allow for the indoor temperature to almost never exceed the outdoor temperature. This is often not attainable with mechanically ventilated greenhouses due to the very large amounts of air that such systems would have to move through the greenhouse to accomplish the same results. When insect screens are installed over ventilation openings, the additional resistance to airflow created by the screen material has to be taken into account to ensure proper ventilation rates. Often, the screen area is made larger compared to the inlet area to allow sufficient amounts of air to enter a greenhouse. The same is true for evaporative cooling pads.Whichever ventilation system is used, uniform air distribution inside the greenhouse is important because uniform crop production is only possible when the plants experience the same environmental conditions.

Horizontal airflow fans

Horizontal airflow fans are frequently installed to ensure proper air mixing. The recommended fan capacity is approximately 3 cubic feet per minute per square foot of growing area. While horizontal airflow fans use a small amount of electricity, they are typically turned off once the ventilation rate exceeds some lower threshold value. Vertical flow of air is also considered important and devices for this are being developed.

Additional cooling or cooling of greenhouse

When the regular ventilation system is unable to provide sufficient cooling for greenhouse temperature control, additional cooling is needed. Two cooling systems, the pad-and-fan and a fog system, are commonly used in greenhouses. Both make use of the cooling effect resulting from the evaporation of water. The process of evaporation requires heat that is removed from the air surrounding the evaporating water.

Pad-and-fan

Pad-and-fan systems can only work in combination with mechanical ventilation. An evaporative cooling pad is installed in the ventilation opening, cooling the incoming air. As the air moves through the greenhouse towards the exhaust fans, it picks up heat from the greenhouse environment. Therefore, pad-and-fan systems experience a temperature gradient between the inlet (pad) and the outlet (fan) side of the greenhouse. This temperature gradient should be minimal to provide all plants with similar temperatures. However, a gradient of 7°F-10°F is not uncommon.The required evaporative pad area depends on the pad thickness. For the typical, vertically mounted 4-inch thick pads, the required area (in square feet) can be calculated by dividing the total greenhouse ventilation fan capacity (in cubic feet per minute) by the number 250 (the recommended air velocity through the pad). For 6-inch thick pads, the fan capacity should be divided by the number 350. The recommended minimum pump capacity is 0.5 and 0.8 gallons per minute per linear foot of pad for the 4- and 6-inch thick pads, respectively. The recommended minimum sump tank capacity is 0.8 and 1 gallon per square foot of pad area for the 4- and 6-inch pads, respectively. For evaporative cooling pads, the estimated maximum water usage is 20-30 gallons per minute per 100 square feet of pad area. Approximately 10 percent of the returning water should be bled off to prevent salt build-up on the pads. Salt buildup reduces the efficiency of the pads. It is recommended that the pads dry out overnight to prevent the growth of algae.

Fog systems

Fog nozzles can be installed throughout the greenhouse, resulting in a more uniform cooling pattern compared to the pad-and-fan system. The recommended spacing is one nozzle for every 50-100 square feet of growing area. The water pressure used in greenhouse fog systems is very high (500 pounds per square inch and higher) in order to produce very fine droplets that evaporate before the droplets can reach plant surfaces. The water usage per nozzle is small, approximately 1-1.2 gallons per hour.The water needs to be free of impurities to prevent clogging of the small nozzle openings. Water treatment and a high-pressure pump are needed to operate a fog system. The usually small diameter supply lines should be able to withstand the high water pressure. Therefore, fog systems can be more expensive to install compared to pad-and-fan systems.

Mist cooling

Evaporative cooling by spraying tiny water droplets into the greenhouse has met with limited success. The droplets must be tiny, and this requires tiny, closely spaced nozzles operated at relatively high pressures — an expensive design.

Water must be well filtered to prevent nozzles from clogging. Uniform distribution of the water droplets throughout the house is difficult to accomplish.If the mist system carries any minerals in the water, deposits will be left on plant foliage. This accumulation can reduce photosynthesis substantially and can lead to salt toxicity. The mist system can also cause wet foliage, leading to disease problems, particularly when the droplet size is too large.Mist cooling does not cool as effectively as a conventional evaporative cooling pad system but it is less expensive. The system requires no collection pan or sump. It can cause runoff or puddling beneath the pads if all the water sprayed on the pads is not vaporized.

Humidity control

Healthy plants can transpire a lot of water, resulting in an increase in the humidity of the greenhouse air. A high relative humidity (above 80-85 percent) should be avoided because it can increase the incidence of disease and reduce plant transpiration. Sufficient venting or successively heating and venting can prevent condensation on plant surfaces and the greenhouse structure. The use of cooling systems (e.g., pad-and-fan or fog) during the warmer summer months increases the greenhouse air humidity.During periods of warm and humid outdoor conditions, humidity control inside a greenhouse can be a challenge. Greenhouses located in dry, dessert environments benefit greatly from evaporative cooling systems because large amounts of water can be evaporated into the incoming air, resulting in significant temperature drops. Since the relative humidity alone does not tell anything about the absolute water holding capacity of the air (the temperature also needs to be known to determine the amount of water the air can hold), a different measurement called the vapour pressure deficit (VPD), is sometime used to describe the absolute moisture status of the air. The vapour pressure deficit is a measure of the difference between the amount of moisture the air contains at a given moment and the amount of moisture it can hold at that temperature when the air would be saturated (i.e., when condensation starts, also known as the dew point temperature). A VPD measurement can indicate how easy it is for plants to transpire. Higher VPD values stimulate transpiration, but too high can cause wilting. Lower VPD values inhibit transpiration and can lead to condensation on leaf and greenhouse surfaces. Typical VPD measurements in greenhouses range between 0 and 1 pound per square inch (0-7 kilopascals).

Passive cooling

Naturally ventilated greenhouses promote cooling in microclimate of the greenhouse. Type of crop amenable for passive cooling greenhouses are selected. Now retractable greenhouses are getting popular to take advantage of natural environment. Obviously passive cooling is cheaper than active cooling.

Shading

Shade curtains help reduce the energy load on a greenhouse crop during warm and sunny conditions and they help reduce heat radiation losses at night. Energy savings of up to 30 percent have been reported, ensuring a quick payback period based on today's fuel prices. Movable curtains can be operated automatically with a motorized system that is controlled by a light sensor and/or timer. Even low-cost greenhouses can benefit from the installation of a shade system.Curtain materials are available in many different configurations from low to high shade percentages depending on the crop requirements and the local solar radiation conditions. In some cases, multiple layers of curtains are installed that can be controlled independently for more fine-tuning.Movable shade curtains can be installed inside or outside (on top or above the glazing) the greenhouse. When shade systems are located in close proximity to heat sources (e.g., unit heaters or carbon dioxide burners), it is a good idea to install a curtain material with a low flammability. The low flammable curtain materials can stop fires from rapidly spreading throughout an entire greenhouse when all the curtains are closed. Shading the greenhouses or using stand-alone shade house in India are common. For cooling plant nursery areas are covered by shade net during summer. During winters shade net protect plants from chill or very low temperature.

Heating of greenhouses

In temperate countries of the wold heating of greenhouses is major and expensive activity. The heating of greenhouses is done using gas, hydropower, solar power or geothermal energy or heat. In India heating of greenhouses is uncommon. A good heating system is one of the most important steps to successful plant production in cold region or during severe winters at other places. Any heating system that provides uniform temperature control without releasing material harmful to the plants is desirable. Suitable energy source is electricity. Savings in labour could justify a more expensive heating system with automatic controls. Most undesirable heat loss from a greenhouse occurs by long-wave radiation, conduction and convection and by infiltration. Heating requirements are determined by calculating the sum of convection and infiltration.

Greenhouse heater requirements depend upon the amount of heat loss from the structure. Heat loss from a greenhouse usually occurs by all three modes of heat transfer: conduction, convection and radiation. Usually many types of heat exchange occur simultaneously. The heat demand for a greenhouse is normally calculated by combining all three losses as a coefficient in a heat loss equation.

Conduction

Heat is conducted either through a substance or between objects by direct physical contact. The rate of conduction between two objects depends on the area, path length, temperature difference and physical properties of the substance(s). Heat transfer by conduction is most easily reduced by replacing a material that conducts heat rapidly with a poor thermal conductor (insulator) or by placing an insulator in the heat flow path. Air is a very poor heat conductor and therefore a good heat insulator.

Convection

Convection heat transfer is the physical movement of a warm gas or liquid to a colder location. Heat losses by convection inside the greenhouse occur through ventilation and infiltration (fans and air leaks). Heat transfer by convection includes not only the movement of air but also the movement of water vapour. When water in the greenhouse evaporates, it absorbs energy. When water vapour condenses back to a liquid, it releases energy. So when water vapour condenses on the surface of an object, it releases energy to the outside environment.

Radiation

Radiation heat transfer occurs between two bodies without direct contact or the need for a medium such as air. Like light, heat radiation follows a straight line and is either reflected, transmitted or absorbed upon striking an object. Radiant energy must be absorbed to be converted to heat. All objects release heat in all directions in the form of radiant energy. The rate of radiation heat transfer varies with the area of an object, and temperature and surface characteristics of the two bodies involved. Radiant heat losses from an object can be reduced by surrounding the object with a highly reflective, opaque barrier. Such a barrier (1) reflects the radiant energy back to its source, (2) absorbs very little radiation so it does not heat up and re-radiate energy to outside objects, and (3) prevents objects from "seeing" each other, a necessary element for radiant energy exchange to occur.

Factors affecting heat loss

Heat loss by air infiltration depends on the age, condition and type of greenhouse. Older greenhouses or those in poor condition generally have cracks around doors or holes in covering material through which large amounts of cold air may enter. Greenhouses covered with large sheets of glazing materials, large sheets of fiberglass, or a single or double layer of rigid or flexible plastic have less infiltration.

The greenhouse ventilation system also has a large effect on infiltration. Inlet and outlet fan shutters often allow a large air exchange if they do not close tightly due to poor design, dirt, damage or lack of lubrication. Window vents seal better than inlet shutters, but even they require maintenance to ensure a tight seal when closed.

Solar radiation enters a greenhouse and is absorbed by plants, soil and greenhouse fixtures. The warm objects then re-radiate this energy outward. The amount of radiant heat loss depends on the type of glazing, ambient temperature and amount of cloud cover. Rigid plastic and glass materials exhibit the "greenhouse effect" because they allow less than 4 percent of the thermal radiation to pass back through to the outside.

Heating Systems

Unit Space Heater

Unit space heaters, either floor mounted or supported, are normally fueled with natural or bottled gas or fuel oil and use fans for heat distribution in western coumtries. This system requires a relatively moderate capital investment, is easy to install, and provides for easy expansion of facilities. These units produce carbon dioxide gas, which is necessary for improved plant growth.

Hot Water Systems

Hot water systems utilizing piping that can be perimeter, under benches, or overhead fan forced unit heaters can be used. These require a boiler, valves, and other necessary controls. However, a hot water system is simpler to install and normally requires less maintenance than a steam system. There are slower heating and cooling of pipes, but temperatures are normally more uniform. Hot water systems are mainly used in smaller ranges. Geothermal water heaters where ever available works out to be cheaper.

Steam Heating System

A steam heating system needs a boiler, valves, traps and other controls depending upon the size and type of boiler used. Steam provides rapid heating and cooling of the steam lines and usually less pipe is needed. Lines may be smooth or finned, and about 1/3 of the heat should be overhead and about 2/3 along the side walls. Lines can also be arranged under benches or with overhead fan-forced unit heaters. A steam system also allows the use of steam for soil pasteurization. A steam system requires a high initial investment; however, it has a long life expectancy. Steam heating systems are most often used in large ranges as steam can be transported long distances efficiently.

Unit Radiant Heaters

Unit radiant heaters using gas of the "Salamander" grove heater (return stack) type and which use fuel oil may be used for frost protection. These are not recommended for routine greenhouse heating because they are not thermostatically controlled making temperature control difficult.

Poly-Tube Systems

Poly-tube or fan-jet systems are frequently used to provide more uniform heat distribution, air movement, and ventilation in greenhouses using any of the above heat source systems.

Precautions

1. The heating system should allow for easy expansion of the range in the future, and the system should have sufficient capacity to offset the heat loss from the greenhouse under the most severe conditions.
2. Normally, design for temperatures slightly above minimum 15 to 25 year lows as shown by local weather records.
2. Provide an adequate system of automatic controls or be prepared for extensive manual control during severe weather. The most important control in most heating systems is the thermostat, which is used to control the operation of the heating system.
3. If possible, select a heating system that will allow you to convert from one fuel source to another. In India, heating of greenhouses is not common because of heavy cost.

23

Greenhouse Equipment, Materials of Construction for Traditional and Low Cost Greenhouses

The variety of greenhouses and greenhouse material, equipment and accessories are so many. However, there are certain basic materials that essential in almost all types of green houses. These are

Structural framing materials

Such materialsinclude low maintenance non-heat conducting timber; strong and durable galvanized steel; or light, low-maintenance and sturdy aluminium alloys. Though plastic pipe frames are usable, they are not very strong to withstand snow load or strong wind sheering forces. What one should choose as the framing material would be determined by the material one wishes to use as glazing, his budget and the permanence of the structure one desires. Aluminium is a popular material for greenhouse framework because it is relatively lightweight, rustproof, affordable and easy to assemble. The downside is that it is not a good insulating material and in very cold weather, heat loss from the metal frame could inflate your heating bill. Also, you would need to look for sturdy frames if your greenhouse is located in an area that experiences high winds. Condensation could also occur on an aluminium frame. Wooden structural frameworks for garden greenhouses are also a popular choice for its aesthetic value, natural feel and the 'warmth' it exudes. It is an easy material to work with especially if you are not using a greenhouse kit for building. Some greenhouse owners would settle for nothing else as they believe it is the only material that complements the natural beauty of plants inside the greenhouse! Wood is a good insulating material so it will not allow heat loss in cold weather. Reinforced wood has superior strength and has an excellent load bearing capacity. Being a natural material, it has the potential to develop mildew or disease and offer a haven to tiny pests. So, when you choose wood to build the framework of your greenhouse, be sure that you pick a high quality hardy wood such as redwood, cypress or cedar. Also, treat wood with non-hazardous chemicals so that your plants are not adversely affected. The preservatives used for wooden greenhouse frames should be pressure treated, salt type or copper naphthenate. Copper naphthenate is a general use preservative used for pressure treating utility poles and timber. It is known to be very effective in preventing mould, mildew and dry rot. It works very well against wood-damaging fungi and insect pests. Pentachlorophenol is a restricted use wood preservative and is toxic to plants and humans. It is not available to the general public and poles or timber treated with it are not suitable for use in a greenhouse. PVC pipe can also be used as support for small greenhouses that are to be covered with polyethylene, but PVC requires additional support to withstand strong winds. Being flexible, PVC can be used particularly well in hoop houses. Though it has the advantage of offering insulation and, if UV treated, are long-lasting, they are not good at bearing the load of hanging baskets and shelving to any appreciable degree. PVC that is treated to be resistant to ultraviolet radiation is also known as UPVC.

Another option for greenhouse framing is Solexx. These greenhouse frames have excellent structural strength being made of combination composite tubing, steel and PVC. The advantage of using these is that while composite tubing and steel provide amazing strength, PVC provides insulation and minimizes condensation on the frame. The Solexx greenhouses have no trouble with harsh winds or weather. In addition, they are also easy to clean.

Covering materials or glazing materialyou pick must be durable, low-maintenance, allowing for best light transmission and least heat loss. You need to keep these factors in mind over and above aesthetic appeal. Covering options use glass or plastic of various types. You have a choice of polycarbonate panels, polyethylene sheets, rigid fiberglass, or plastic films – each with its unique advantage or disadvantage. You would need to match the framing material with your choice of covering. Generally, double glazing gives a good result with heat retention. During winters when natural light is reduced, there is a direct analogous reduction in plant growth. So, a covering material that has good transmission properties is a desirable choice. Common glazing materials transmit about 80-90 per cent of the light that is incident on them. Materials such as glass allow for direct transmission of light while other materials allow diffuse light transmission. The diffuse light creates a more even distribution of light than does direct light which is characterized by sharp areas of illumination and shadow.

Glass is the traditional covering material for greenhouses. Glass used for greenhouses today is tempered to augment its strength. Glass is great for direct light transmission and durability. The tinted varieties reduce transmission of light significantly, though they correspondingly reduce heat loss as well. Glass may prove to be an economical choice in the long run, but initial installation costs are significant. Also, though tempered glass is strong, it is not completely hail-proof and there is a risk of shattering.

Glass is traditionally used in greenhouses for its excellent light transmission and though it allows about 94 per cent light to go through, being a bad heat conductor it retains heat and can cause rapid heating of the internal environment. This is good news in very cold winters, but it could ruin your plants in hot summers. So, ventilation and temperature control become critical factors in this regard.

In addition, glass is a lot heavier than some of the other glazing materials available and so, you would need a correspondingly sturdy framework to hold it up securely. Generally, heavy frameworks translate into heavy investments too. You have to be extra careful in transporting glass as it is fragile in this regard. This is not to discourage you completely from opting for glass, it does have its advantages.

Its light transmission in excellent, no other glazing is as durable and glass in the long run, and if you can avoid breakage, there is hardly any maintenance needed. As it comes as panels of convenient sizes, you would need to replace only the individual damaged panels without any effect to the rest of the glazing. Plus, the aesthetic value of glass is incomparable. Your greenhouse would look as beautiful from outside as inside!

Polycarbonate is an increasingly popular choice for greenhouse coverings due to its many desirable features. It is very strong with high impact value, hail-proof, clear to allow excellent transmission of light, reasonably flexible so that it can be bent, and is less flammable than other plastic materials. It comes as single, double or triple-wall structured sheet. The single wall polycarbonate sheets are available as corrugated sheets in matte and clear varieties. The polycarbonate rolls available for greenhouses already come with a coating of material that reduces dripping of moisture.

Polycarbonates tend to expand or contract with temperature changes and installation must be accordingly done to allow for this feature. Single sheet variety can be installed with screws and rubber gaskets, while the multiple-layered sheets can be secured to the frame with aluminium extrusions.

Aluminium tape is used to seal the top and bottom of the sheet to prevent entry of dust, insects or algae. Polycarbonate sheets score over plastic films in their estimated life span which is around 15 years or more as compared to plastic films for which it is around 4 years. Glass, of course, outlives both at a life span of about 25 years.

Plastic films are good options especially for hoop houses as they are very flexible. They are available in varying grades of quality. Plastic films are low cost, flexible and lightweight. If you opt for these films; your overall structural costs go down because you need lighter frames to support these. Woven polyethylene films are extremely resistant to tears. Sheets designed especially for use as coverings of greenhouses are UV stabilized, have light transmission comparable to glass and last for about 4 years. Once properly attached to the frame they can withstand a good deal of the shearing force of the wind as well as heavy snow or hail. Another attractive feature is their additional ability to reduce heat loss due to the presence of infrared inhibitors that keep heat inside. The plastic films are made with anti-drip surfaces that prevent condensation of moisture on the film. The 6mm UV stabilized film is standard for greenhouse use, though you will find films of various thicknesses in the market. Polyethylene is the most inexpensive covering for a greenhouse.

The plastic films include polyethylene films, PVC or polyvinyl chloride sheets and copolymers. Polyethylene lasts much less than the PVC or copolymers which last up to five years. These usually cost much more than polyethylene. UPVC is a rigid plastic material that was first used in the construction industry. UPVC has the potential to be used as glazing as well. It comes as standard units of double glazing, as custom made prefabricated panels that are brought to the site ready to be simply bolted together. Ordinary utility polyethylene that is not UV stabilized can be used as glazing too, but if you are going to build a greenhouse glazing is not one of the areas you should be economizing on. Utility polyethylene would last you about a year. Glazing is critical to the performance of a greenhouse, so you should opt for one that is specifically designed for use in greenhouses. Earlier, polyethylene sheets were too fragile, prone to easy tearing and damage by the rays of the sun. Not only this, they also allowed condensation to occur on the inside walls of the greenhouse which attracted dirt, grime and resulted in diseases breaking out in the greenhouse. The polyethylene available for greenhouses addresses all these issues, but it is still the glazing that requires most frequent changing. Being the cheapest option in glazing, polyethylene is an attractive choice, but do study other glazing materials before you zero in on this one.

Other than the coverings discussed above, fiberglass and acrylic coverings are also available, though these are losing popularity and are being increasingly replaced by polycarbonates or plastic films. Fiberglass is treated to have a coating of plastic resin over glass to create fiberglass reinforced plastic or FRP. This coating tends to erode with the passage of time – in about seven to ten years - leaving the underlying glass exposed to dirt, grime and dust. Accumulation of these significantly affects light transmission. You could get the covering recoated, but that would involve the laborious process of washing and cleaning the discolored sheet and then overlaying it with a single layer of plastic resin by painting.

Acrylic sheets that are double-layered and used as coverings are longer lasting than polycarbonates. However, they tend to yellow with age thereby reducing light transmission. Some disadvantage with these is that they are flammable, cost more than polycarbonate sheets and are not hail-proof. Acrylic sheets lose to polycarbonates in other areas too – they are easily scratched and gather grime and dust more easily. In addition, these sheets are not as flexible as polycarbonates and therefore are generally used on flat straight surfaces; the method of securing them being quite the same as in case of the polycarbonate sheets. They too need to be secured allowing scope for expansion.

Shade net/cloth

Shade net or cloth is very common among plant nurserymen in the country. It does two things very good.

1. It reduces harmful sun radiation, thereby increasing photosynthetic active radiation (PAR) which is really what we want.
2. It protects the crop against hail and wind.

The agro shade nets are available which easily last 3 to 5 years depending on the climatic condition. They are available in different colours with wide range of shade percentages viz. 25%, 30%, 35%, 50%, 60%, 75% and 90%.

Shade net house are considered as one of the well adopted technologies to provide development of healthy grafts/ seedlings and hardening for various horticultural crops irrespective of climatic conditions. Heating is not viable for obvious reasons but during winter particularly night time it provides protection against frost and low temperature.

Insect screen

Insect screen is used to mechanically discourage the insects to enter the green house and helps in saving on chemical spaying and polluting the environment besides the danger residual pesticides in the produce is not there.

Screens reduce solar radiation and air velocity by about 15–39% and 50–87%, respectively; increase air relative humidity by 2–21%; decrease air temperature and evapotranspiration by 2.3–2.5 °C and 17.4–50% respectively.

Screens with varied mesh (number of holes per inch) i. e. 17-55 are commercially available. They are used keeping in mind the insects of the area and climate. 50 mesh net is used for blocking whiteflies, aphids and leaf miner. Screens are generally white but can be coloured also depending on the technical requirement.

Addition of UV-stabilizers, IR-absorbers to the claddings and the development of the anti-fog treatments, anti -drip and others as per requirement of the specific region is done.

Types of additives to claddings

1. UV (290-400 nm) absorbers and stabilisers increase durability, reduce the potential damage to biological systems in the greenhouse and may control some plant pathogens
2. Infrared (700-2500 nm) absorbers reduce long wave radiation and minimise heat loss
3. Long wave radiation (2500-40000 nm) absorbers reduce the loss of heat radiated from materials and objects (including plants) inside the greenhouse
4. Light diffusers scatter light entering the greenhouse, reducing the risk of plants getting burnt and improving the amount of light available to the lower parts of the plant.
5. Surfactants reduce the surface tension of water, dispersing condensation.
6. Antistatic agents reduce the tendency of dust to accumulate on plastic films.

Besides

1. Colour pigments may improve plant growth by altering the proportion of selected wavelength ranges
2. Fluorescence may be used to increase the emission of red light
3. Glossy surfaces may repel insects

The process of making multilayer films enables thin layers of materials with different properties to be joined to make superior composite films. Properties such as durability, creep (deformation over time) and long wave radiation absorption are being be improved.

24

Cost Estimation and Economic Analysis

There are different types of greenhouses, the low cost and high cost or low technology or high technology greenhouses. The cost of greenhouse and expenditure on crop production under these greenhouses would vary according to technology to be used and the crop is to be produced. The cost would vary from place to place with location of greenhouse. For each type of greenhouse its cost would vary according to geo-climate, the production practices and input expenditure/availability. The cost of construction of greenhouse in 2019 hence varies from Rupees 500 to 6000 per square meter.

Table: Types of greenhouses and their cost

S.No.	Types of greenhouse	Cost (Rupees) per square meter
1.	Low tech. greenhouse without fan and pad	500
2.	Medium tech. greenhouse with fan and pad system without automation	1000-1200
3.	High tech. greenhouses with automation	2500-6000

National Horticulture Board (NHB), Ministry of Agriculture and Farmer's Welfare, Government of India did following costing of greenhouse structure for subsidy

A. Fan & Pad System: Rs 1400/Sqm and Rs 1610/Sqm for hilly areas

B. Naturally Ventilated System

1. Tubular Structure Rs 844/sq. and 970/sqm for hilly areas
2. Wooden structure Rs 540/sqm and Rs 621/sqm for hilly areas
3. Bamboo structure Rs 450/sqm and Rs 518/sqm for hilly areas

The rate per square per meter would change with the quality of material and further advancement in the technology.

Economic Analysis

Greenhouse enterprise budget is comprised of two types of costs, variable and fixed. Variable costs are those costs which are incurred only if the production cycle is started. Seeds, fertilizer and pesticides are examples of such costs. Fixed costs occur whether or not there is production. Local taxes, insurance, depreciation and interest on investment such as structure and equipment are examples of fixed costs and must be accounted for even if there is no production. An economic analysis of greenhouse enterprise budget can be used to estimate the profitability of it by including sales revenue and net returns. Net returns are expressed as gross margin and net income. Gross margin is expressed as revenue minus variable cost and net income is revenue minus all costs. Unpaid family labour is charged to the enterprise as an expense because it represents the loss of opportunity for the family member to work elsewhere and earn income.

Greenhouse production costs and returns computation

The following components on investments and returns are generally taken into account.

Interest on Investment

$$\text{Depreciation} = \frac{\text{Purchase Price- Salvage value}}{\text{Number of years of life}}$$

Land Value

Property or business tax if any

Labour costs

Production material and supplies

Cooling/heating cost

Utility cost

Transportation cost

Repair and maintenance cost

Marketing charges - grading, packaging, marketing and administrative fees

Miscellaneous costs- other costs incurred in a greenhouse operation, but not reported under any other heading

Cost Analysis

Net Present Worth (NPW)

The NPW is defined as the difference between present worth of savings and cost of investment. The mathematical statement for net present worth can be written as

$$\text{NPW} = \sum_{t=1}^{t=n} \frac{B_t - C_t}{(1+i)^t}$$

Where

C_t = cost in each year

B_t = Benefit in each year

t = 1,2,3 ---------- n

i = discount rate

Internal Rate of Return

The internal rate of return is threshold rate at which the NPW is zero. Internal rate of return is discount rate(i) such that

$$\sum_{t=1}^{t=n} \frac{B_t - C_t}{(1+i)^t} = 0$$

Benefit Cost Ratio

The ratio is obtained when the present worth of the benefit stream is divided by the present worth of the cost return. The mathematical benefit-cost ratio can be expressed as

$$\text{Benefit-cost ratio} = \frac{\sum_{t=1}^{t=n} \frac{Bt}{(1+i)^t}}{\sum_{t=1}^{t=n} \frac{Ct}{(1+i)^t}}$$

Payback Period

The payback period is the length of time from the beginning of the project until the net value of the incremental production stream reaches the total amount of the capital investment. It shows the length of time between cumulative net cash flow recovered in the form of yearly net cash flows.

$$PP = \frac{IC}{ANI}$$

Where

PP= Payback load

IC= Investment Cost

ANC = Annual Net Income

Break- Even Analysis

Break-even analysis presents the point where there is just sufficient revenue to cover the costs. It is the point at which the total cost and the total gross revenue intersect. . It is a method used more frequently to demonstrate the probable effects of change than to determine what those changes should be

$$BEP = \frac{TFC}{S_P - U_P}$$

Where

BEP=Break-Even Point; the volume $T_r = T_c$

TFC=total fixed cost per year

$S_{p=}$ selling price per Kg

Up= $_{Cost}$ per Kg

$_{=}$TVC/(Total wt. /year)

A. Cost Economics of High Cost Greenhouse

High cost greenhouse may be multi-span structure. The cost estimates are indicative and may vary according to location, crop, cladding material and environment control system. The additional cost involved per square meter is given below.

Sl.No	Specifications	Cost Rs/m2
1.	If double layer polyethylene used	100
2.	Co2 generation & distribution	150
3.	Evaporative cooling	200
4	Heating system	100
5.	Humidification system	100
6.	Lighting	200

Contd.

7.	Night curtain / Shading system	150
8.	Drip system	20
9.	Nutrient application system	100
10	Porous flooring	100
11.	Benches	150
12.	Structural cost	300
13.	Miscellaneous	180

Average cost of High Cost Greenhouse per square metre Rs. 2000.00

Estimates of material requirement for simple pipe framed low cost greenhouse (4x25m)

Sno.	Detail of material	Requirements
1.	G.I. Pipe of 25 mm. ö for foundation	50 x 1.20 =60m
2.	G.I. Pipe of 15 mm. ö for arches and top M.S Flates 19x8m	25 x 6 = 150m 25 x 1 = 25m
3.	For sides for holding mesh and UV film and front & backside &for purlins	120.00m run
4.	M.S. 'L' angle19x19x3m for front and end frames and door	28m run
5.	Plastic beading 25 mm width	150 m
6.	UV stabilised film (200 micron)	175 m2
7.	Mesh 40-60 size all round (1.15m width)	65 m
8.	Door frame and fitting	L.S
9.	Hardware like Bolts and nuts, welding rods, Aluminium oxide, Painting etc.	L.S
10.	Earth work and concreting of foundation with 1: 3: 6	L.S
11.	Labour cost	L.S

Cost Economics of Simple, pipe framed greenhouse (100 m2)

Sl.No	Item of work	Qnty. (Rs)	Rate / unit (Rs)	Total amount
1.	G.I. Pipe 25 mm dia.'B Class	60.0m	65.00/=	3900.00
2.	G.I. Pipe 15 mm dia. 'B Class	175.0m	25.00/=	4375.00
3.	M.S.Flat 19x3mm Size	80.0kg	16.50/kg	1320.00
4.	M.S.L angle 19x19x3mm Size	20.0kg	17.50/kg	350.00
5.	UV stabilised film	175.0m2	20.00/=	3500.00
6.	Plastic beading	150.0m	4.00/=	600.00
7.	Mesh 40-60 size all round	65.0m2	22.00/=	1430.00
8.	Door frame and fitting	L.S	425.00	
9.	Labour cost	L.S	650.00	
10.	Bolts, nuts and welding rods	L.S	400.00	
11.	Earth work and concreting of foundation	L.S	550.00	
	Total	17,500.00		

Average cost per square metre Rs.175.00

Cost Estimates of Medium Cost Greenhouse (4 x 25 m2)

Sl.No	Specification	Cost (Rs.)
1.	Cost of Greenhouse (100 m2)	17,5000.00
2.	Additional items on GH-2 (for fans –2)	1000.00
3.	Electrical Fittings (Power point distribution, boxes, MCB etc)	3000.00
4.	Mist spray assembly	2000.00
5.	Inflation blower	1000.00
6.	Two fans (60 Cm dia)	10000.00
7.	Monoblock AC pump (3hp)	3500.00
8.	Water tank (Sintex) 1000 liters	4000.00
9.	Thermostat/ Humidstat	2000.00
10.	Cooling pad and fittings (local made)	3000.00
11.	Labour (extra)	1500.00
12.	Additional film for double layer	3500.00
	Total expense	50,000.00

Cost of greenhouse of medium type per square metre Rs.525.00

Detailed estimate of material requirement for construction of low cost greenhouse (27 x 18.5m = 500m2)

Sl.No	Item of work / Detail of material	Requirement, Qty
A. Structural material		
1.	G.I. Pipe of 56 mm. ö for foundation	42.00m
2.	G.I. Pipe of 50 mm. ö for foundation posts	120.00m
3.	M.S. 'L' angle25x25x6m for arches + top & bottom frame to fix mesh and film + for tie beams(102 lengths)	612.00m
4.	Plastic beading 25 mm width	200.00m
5.	Plastic pipes 25mm ö and 19mm ö as grippers.	100.00m. (each)
6.	Base plate for foundation post (50 x 50 x 6mm0	40.0 nos.
7.	Welding rods	5 boxes
8.	Hardware like Bolts and nuts, welding rods, Aluminium oxide, Painting etc.	L.S
B. Cladding material		
1.	UV stabilised film (200 Micron)	5.00 Bundles
2.	Rambonet 40-60 mesh (1.15m width x 30.0m length	12.00 Bundles
C. Construction material		
1.	Earth work excavation for foundation	4.00 m3
2.	Sand filling for foundation and basement(optimal)	1.00 m3
3.	Plain cement concrete for foundation	2.25 m3
4.	Hollow cement bricks for foundation	3000.00nos

Detailed estimate of material requirement for construction of medium cost (single span) greenhouse (33 x 9.1m = 300m2)

Sl.No.	Item of work / Detail of material	Requirement (Qty)
A Structural material		
1.	G.I. Pipe of 56 mm. φ for foundation	30.00m
2.	G.I. Pipe of 50 mm. φ for foundation posts	90.00m
	G.I. Pipe of 38 mm. φ for purlins & arches	205.00m
3.	M.S. 'L' angle 25x25x3m for top & bottom frame & door	150.00m
4.	Plastic beading 25 mm width	200.00m
5.	Plastic pipes 25mm φ and 19mm φ as grippers.	70.00m. each
6.	Base plate for foundation post (50 x 50 x 6mm φ	25.0 nos.
7.	Welding rods	3 boxes
8.	Hardware like Bolts and nuts, welding rods, Aluminium oxide, Painting etc.	L.S
B Cladding material		
1.	UV stabilised film (200 Micron)	150.00Kg
C Cooling system (fan 7 pad)		
1.	Fans (air flow fans)	3.00 Nos
2.	Pad (9.0m x 1.5 m)	1.0 Nos
D Construction material		
1.	Earth work excavation for foundation	1.25 m3
2.	Plain cement concrete for foundation (1:2:4)	1.25 m3
4.	Hollow cement bricks for parapet wall	-1000 no

All figures are indicative

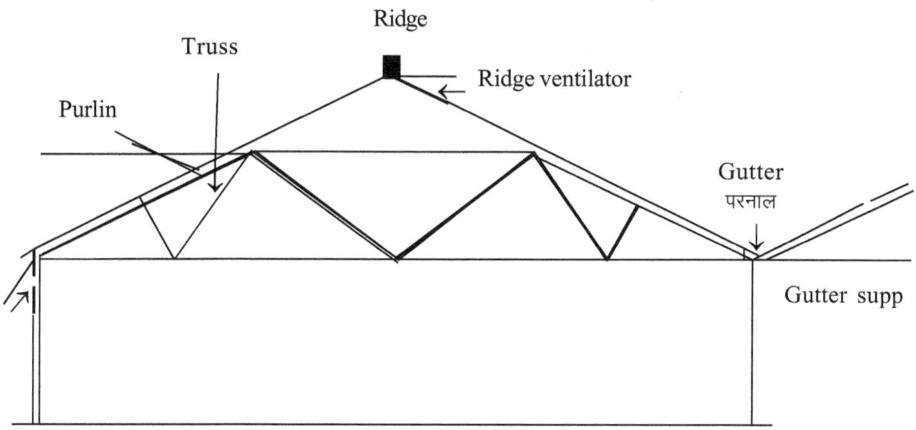

Lateral ventilator

25

Passive Solar Greenhouse

The modern or climate controlled greenhouse requires massive inputs of energy and cost of claddings to grow crops out of season making them high cost structures. However, growing of certain horticulture crops and their nurseries out of season is happening in a sustainable way, using the heat from the sun and soil too. These are called passive solar greenhouses designed to retain as much warmth as possible. This concept has revolutionized vegetable production in Ladakh, high altitudes and other parts of the country in winter months.

The solar greenhouse is especially successful in China and are contributing mainly in vegetable production. Solar passive greenhouses are designed in different shape size using different local and other materials keeping in view the local climate and terrain. The poly houses mentioned earlier such as, low cost polyhoses, Ladakhi poly house, solar trench, polyench are passive solar green houses. Being the opposite of the climate controlled greenhouse, the passive solar greenhouse is heated all-year round with solar energy alone, even when the outdoor temperature drops below freezing point. The indoor temperature of the structure can be higher than the outdoor temperature. The temperature can be around 25°C (45°F).

Such greenhouses have been experimented by Defence Research and Development Organization through its establishment FRL (Field Research Laboratory) now DIHAR (Defence Institute of High Altitude Research), Leh/ Ladakh since 1962 during November to the end of March, the period during which the outside temperature drops there below freezing. The average temperature in the coldest month is between -15°C and -18°C. Over the several decades of research different and effective designs of such greenhouses have been standardized which are suitable to produce fresh vegetables like most of the leafy vegetables, tomato, cole crops and potato at sub zero temperature.

Plastic Low Tunnels or Row Covers

These are miniature structures producing greenhouse like effect. Plastic low tunnel facilitates entrapment of carbon dioxide thereby enhancing the photosynthetic activity. It protects plants from harsh climatic conditions such as rain, wind, hail, snow, low temperature/frost, etc. These are mainly used for raising nursery for early germination of seeds and production of healthy seedlings. Besides protecting plants from hostile climate these are helpful in round the year cultivation of certain crops. These tunnels are helpful in raising early cucurbits in north India by protecting them from low temperature and frost. Like plastic mulch the cost of low tunnels is affordable by most of the farmers. Besides, they can be used to initially cover the crops for growth to save from abiotic and biotic stress and later expose to favorable natural environment for fruiting by removing cover. Plastic mulch and drip irrigation/ fertigation under poly tunnels are beneficial to check weeds and conserve water better. Non- woven fabrics tunnels are likely to replace plastic tunnels in immediate future.

Non-Woven Fabric for low tunnel- Biodegradable. Easy to handle. Water condensation problem observed in plastic tunnel is not there. Crop is protected from low temperature.

Plastic low tunnel or crop cover

Low tunnel technology offers very good opportunity for early harvesting of cucurbits like, watermelon, muskmelon, long melon, bottle gourd, ridge gourd, round gourd, summer squash, cucumber, *etc.* in hot arid region. The technology should essentially be used with drip irrigation, fertigation, spray of micronutrients and integrated crop management practices to harvest high yield and quality produce. Now a day's cultivation of cucurbits under low tunnels has become reality and emerged as a viable source of income among the farmers of arid region. An advancement of 30-50 days has been observed in harvesting over their normal season of cultivation which provide very high price in the market.

What is low tunnel?

Low tunnels are flexible transparent covering that are installed over the rows or individual beds to enhance plant growth by warming the air around the plants using heat from the sun especially during winter season. Plastic tunnels are transparent which provides required sunshine to the plants, and the plastic also plays a barrier against the cool air in winter.

Why low tunnels?

Presently, river bed cultivation is in practice for production of cucurbitaceous vegetables in off-season in northern parts of India, although area under river bed cultivation is very limited, which cannot be extended further. Besides, glut in the market and low quality produce causes poor economic returns to the farmers. But with the use of low tunnels, cucurbits can be grown very early in the spring or summer season so that the produce to the market can be sent early in the season. It also extends the growing season for selected vegetable crops when large quantities of the crop produce are not available results in higher prices from their off-season produce.

Principle of low tunnels

In our surrounding atmosphere CO_2 concentration is 0.03% means 300 ppm. Plants use this CO_2 for photosynthesis. In poly tunnel, during night time there is no photosynthesis but CO_2 is given out through respiration. This CO_2 remain accumulated around plants hence the CO_2 concentration is higher inside the poly tunnels as compared to outside and this CO_2 is again used by plants growing in ploy tunnels for rapid photosynthesis.

Advantages

i. Early and off-season production of cucurbits can be taken to get better return.

ii. Makes cultivation of vegetables possible in areas where it can't grow in open conditions *viz*., high altitudes and hot arid regions.

iii. It creates optimum microclimate for plant growth, thus, increases photosynthetic activities and thereby yield.

iv. Maintains optimum temperature for plant growth.

v. Enhances nutrients uptake by the plants.

vi. Provide protection against unfavourable environment like high rainfall, hail, low temperature, frost, wind, insect-pests, *etc*.

vii. Inexpensive, as these structures are easy to construct and dismantle.

viii. It generates better revenue even though total yield may be the same or lower because market prices are higher.

Components

i. Structural	:	GI wires (2.0 m long with 4-6 mm diameter) or *phalsa* twigs or bamboo sticks.
ii. Covering material	:	Transparent and biodegradable polyethylene film of 30-50 microns having 2 meter width.

Important points for tunnel cultivation

i. Prior to start off-season vegetable cultivation in tunnels, the farmer must have practical knowledge about vegetable cultivation.

ii. Soil and water quality should be tested before starting the cultivation.

iii. Recommended package and practices should be followed.

iv. Farmer must have the updated market information to earn high profit.

Table: Recommended improved varieties and seed rate

S.No.	Crop	Varieties	Seed rate (kg/ha)
1.	Bottle gourd	Thar Samridhi, Pusa Naveen, Pusa Summer Prolific Long, Pusa Summer Prolific Round, Pusa Sandesh, Pusa Santushti.	2.5-3.0
2.	Longmelon	Thar Sheetal, Punjab Longmelon-1	1.5-2.0
3.	Muskmelon	Durgapura Madhu, Hara Madhu, Pusa Madhuras, Pusa Madhurima, Punjab Sunheri.	2.0-2.5
4.	Ridge gourd	Thar Karni, Pusa Nasdar, Pusa Nutan.	2.0-2.5
5.	Summer squash	Pusa Alankar, Early Yellow Prolific, Australian Green, Pusa Pasand, Punjab Chappan Kadddu-1.	3.0-4.0
6.	*Tinda*	Pusa Raunak, S-48.	2.5-3.0
7.	Watermelon	Thar Manak, Sugar Baby, Durgapura Lal.	2.0-3.0

Field preparations and sowing

The land should be prepared to a fine tilth before construction of low tunnels. Application of manures and fertilizers depends upon the crop, variety and soil status. Well decomposed FYM should be applied @ 150-200 q/ ha, phosphorus @ 60-80 kg/ ha and potash @ 50-60 kg/ ha at the time of final land preparation. Nitrogen should be applied @ 60-80 kg/ ha in three split doses. One-third of N should be applied basally, about 8-10 cm away from the seeds. Remaining dose of nitrogen should be given at the time of earthling up and initiation of flowering as topdressing. 45-60 cm deep and 45-60 cm wide trenches should be made at a distance of 2.0-2.5 m in east-west direction in December end. Recommended doses of fertilizers (NPK) should be applied in trenches and mixed thoroughly. The excess dose of nitrogenous fertilizers should be avoided as it results in more vegetative growth and less pistillate flowers. For irrigation one lateral (12-16 mm size) in each trench having drippers of 4 litre/ hour discharge spaced at 60 cm distance should be placed.

Before sowing, the seeds should be allowed to soak in water for 3-4 hours (muskmelon, long melon, cucumber), 6-8 hours (bottle gourd), 10-12 hours (*tinda*, watermelon, ridge gourd) and 24-36 hours (bitter gourd) for quick germination. After soaking, seeds should be treated with Captan or Thiram @ 2 g/ kg, wrapped in gunny bag and kept at warm place (straw) for 2-3 days to facilitate early germination. Daily monitoring of seed in gunny bag should be done to avoid any damage. Sowing of pre-germinated seeds should be done from third week of December to first week of January. Two seeds near each dripper should be done to maintain optimum plant density. The trenches should be irrigated with drip irrigation prior to sowing.

Installation of low tunnels

Low tunnels are constructed with bamboo or iron rods to form semi-circular frames, which are then covered with transparent plastic sheet or white coloured non-woven cloth. Such structures are temporary and usually no higher than 1.0 m. Just after sowing, the flexible galvanized iron hoops (4-6 mm thick) are fixed manually at a distance of 3-4 m on trenches. The width of two ends of hoop and height from ground level should be kept 1.0 m. The erected iron hoops should be covered with transparent 30-50micron biodegradable plastic sheet of 2.0 m width on the day of sowing. Both the perpendicular ends of plastic sheet should be buried in the soil to make low tunnel. It is done manually however; tractor driven machines are also available in advanced countries for this purpose. The prepared tunnel reflects infra-red radiation and keep inside temperature 8-10°C higher than outside. Locally available material such as *phalsa* twigs can also be used instead of GI wire to reduce the cost.

The increase in temperature inside low tunnel facilitates early germination of seed and crop growth. The plastic sheet serves two purposes: first it traps heat and reduces water loss and second it protects plants from adverse climatic conditions. Good cross ventilation and potential stresses caused by heavy wind, hail or heavy rains must be considered while constructing the structure.

Generally biodegradable plastic film of 30-50micron thickness is used. However, plastic film of 30 micron performed best under arid region. This biodegradable plastic is available according to the requirement of the duration one want to cover the crop or use as mulch in the crop. After that period, the plastic after receiving sufficient sunlight, it becomes brittle. The film eventually breaks down into small flakes and finally completely composted in the soil. The plastic is having vented or silted during the growing season as the temperature increase within the tunnels during peak day time. Vents also help bees to enter inside tunnels for pollination. Generally, 3-4 cm size vents can be made on eastern side of the tunnels just below the top on a distance of 2.5 to 3.0 m. Now a day's non-woven cloth is also being used as covering material instead of plastic which is cheaper than plastic.

Off season crop

Fertigation, irrigation and inter-culture

It is very difficult to obtain full benefit from low tunnels without micro-irrigation and fertigation. Drip irrigation combined with fertigation is essential for growing plants inside the tunnels. It not only applies water in the root zone but also keeps the humidity low leading to less pest and disease problems. Fertigation with water soluble NPK (19:19:19) @ 8-10 kg/ ha should be done at vine development and flowering stages. Spray of 25 ppm boric acid along with 1% urea as the adjuvant 3 times from 8-leaf stage to 45 days after sowing was found beneficial which increase number of pistillate flowers, fruit setting and improve quality of fruits in all cucurbits. Addition of urea as an adjuvant @ 1% to the spray solution improved the absorption of boron by leaves. Optimum soil moisture should be maintained by operating drip system at 5-6 days' interval during December to February) and at an interval of 2-3 days during March-April months for 1-1.5 hour.

Weeding and hoeing should be done along and between the rows. It should be done at the time of topdressing of nitrogenous fertilizer which is generally done before emergence of tendrils. Once the foliage has covered the soil, it is better to stop hoeing since it may damage the roots. Normally two to three hoeing and weedings are required to keep the crop weed free.

Hardening process

When plants start to produce pistillate flowers, the plastic should be removed partially during day time and covered during night hours. In second or third week of February when outside temperature rises, the plastic is completely removed from the plants. While removing the plastic care should be taken that it should not be removed suddenly. Hardening of plants is essential to prevent death of plants. Always remove the plastic during morning hours and cover in evening hours. Repeat this process for 2-3 days to harden the plants and avoid shock.

Pollination

Poor pollination is a major problem at high temperature. Inadequate pollination cause drying of ovary, misshapen and undersize fruit. Pollination is an important factor in cucurbits to be taken care of for good fruit setting thereby increasing total yield. The sex form in most of the cucurbits is monoecious (separate male and female flowers on same plant) hence effective cross pollination is needed. It can be performed by honeybees (*Apis melifera*) which can work in tunnels easily through the vents, made on the plastic. Optimally, one beehive having 30000-50000 workers is sufficient for one-acre area for effective pollination in cucurbits. It is recommended to keep the beehive box on the north-west side of the field for effective working of the bees. Plastic can also be removed during day time when plants start to produce pistillate flowers to facilitate pollination of the crop by visit of bees. It is very important to ensure pollination when there is complete flowering in the plants inside the tunnels because the yield would be reduced considerably if there is poor pollination.

Care of crop during summer

With the onset of summer season in hot arid regions the wind velocity increases which increase the soil temperature, collect the vines in rows leading to poor pollination and flower drop. To overcome this situation, *Saccharum* (*sarkanda*) should be put in the space left between rows to facilitate proper vine spread and to avoid the direct contact of vines with warm soil. Proper direction to the growing vines should be given manually so that each and every pistillate flower gets pollinated resulting in increased yield.

Harvesting and crop advancement

Different cucurbits sown during December (third to fourth week) advanced the crop by 30-50 days over their normal season of cultivation. If the longmelon crop sown in third week of December, can be harvested in last week of February or first week of March. Similarly, other cucurbitaceous crops such as muskmelon, watermelon, bottle gourd, round melon, summer squash and ridge gourd can be advanced 30-50 days early than the normal season under low tunnels.

Table: Sowing time, harvesting time and crop advancement under low tunnels

Crop	Sowing time	Harvesting time	Advancement in harvesting Days
Bottle gourd	Third to fourth week of December	Second to third week of March	40-50
Longmelon	Third week of December to first week of January	Last week of February or first week of March	30-50
Muskmelon	Third week of December to first week of January	Second week of April to last week of April	30-40
Ridge gourd	Third to fourth week of December	Second to third week of March	40-50
Summer squash	Third week of December to first week of January	Last week of February	40-50
Tinda	Third to fourth week of December	Last week of February	40-50
Watermelon	Third week of December to first week of January	Second week of April to last week of April	30-40

Crops ready for harvesting

Off season cucurbits produced under low tunnels can fetch very high price in the market. On an average the cost benefit ratio of 2.0-2.75 can be obtained under hot arid conditions.

Table: Yield and expected B:C ratio of under low tunnels in hot arid region

Crop	Cost of cultivation/ ha (Rs.)	Production (q/ ha)	Rate of produce (Rs./ kg)	Gross income /ha(Rs.)	Net income (Rs.)	B: C ratio
Longmelon	90000	180	15	270000	180000	2.00
Muskmelon	80000	150	20	300000	220000	2.75
Watermelon	80000	240	10	240000	160000	2.00

Plant protection measures

Fruit fly, Hadda beetle, leaf eating caterpillar, leaf miner, white fly, aphid and red pumpkin beetle are important insects of cucurbits. Integrated pest management (IPM) practices should be followed to reduce load of insecticides. Soil should be exposed to sun during summer by deep ploughing to kill hibernating pupa of insects. Spray Dimethoate 30EC (2.0ml/ litre water) or Spinosad 45SC (0.5-0.7ml/ litre water) to control fruit fly, Hadda beetle and leaf eating caterpillar. Commercially available Cue-lure traps 7-8 in one-hectare area should be installed to manage fruit fly. Aphid, white fly and leaf miner can be controlled by spraying either Imidacloprid 17.8SL or Thiamethoxam 70WS @ 0.3-0.5ml/ litre water.

Diseases such as downy mildew, *Alternaria* leaf blight, *Fusarium* wilt and mosaic affect the cucurbits. Seed treatment with Captan or Thiram or Bavistin @ 2g/ kg seed should be done prior to sowing. Downy mildew and *Alternaria* leaf blight can be controlled with periodic spray of Indofil M-45 @ 2g/ litre water. Drenching of crop with Bavisitn @ 2g/ litre water should be made to control *Fusarium* wilt. Spry of Imidacloprid 17.8SL @ 0.3-0.5ml/ litre water should be done to manage viral diseases transmitted through aphid and white fly.

Walk –in Tunnels

Walk in tunnels are basically the protected structure made from pre-galvanized pipe and covered with UV stabilized plastic Sheet. Generally, walk in tunnels are used to give favourable environment like temperature, humidity, light intensity, ventilation, soil media, disease control, irrigation, fertigation and other agronomical practices during the season irrespective of the natural conditions outside. Walk in tunnel have variety of applications, the majority being, growing of vegetables, floriculture, planting material acclimatization, fruit crop growing for export market. Walk in tunnels are available in different sizes and constructed as per customer requirement. Walk-in tunnels are mostly covered with UV grade film on G I pipes or other support. Optimum size of walk in tunnel is 120-125 Sq.m. with a width of 4 m, 25-30 m length and 2m central height. These are designed to withstand wind up to 120km/h, and trellising loads up to 25 kg/sq. m. Walk in tunnel structure in gable configuration can also be 8 or 10 meters wide. Walk in tunnels are erected over the crop during the peak winter month (December, January and February) in National Capital Region. Thereafter the cladding is removed as and when temperature is favourable to the crop.

These are suitable for nursery of flowers and vegetables.

Besides raising plant nursery these are suitable for production of tomato, brinjal, broccoli, capsicum, parthenocarpic cucumbers, leafy vegetables, strawberry, Chive, early production of certain vegetables and other crops.

Walk-in tunnels

High Roof Tunnels with Ventilation

High tunnels, also called hoop houses or high hoops, are permanent or temporary structures are fabricated on the site. These are made with galvanised steel pipes and profiles to provide long life to the structure. These are provided with 1.0 m top ventilation along with a central height of approximately 4.7-5.0 m and side ventilation 2.2-2.5 m. Like walk in tunnels this structure also extends the growing season can be used for a variety of vegetables, small fruits and flowers. They are cheaper than most greenhouses. These are suitable for cucumber, capsicum, tomato, gerbera and carnation production.

Carnation crop in high roof tunnel made by using material

Some of the advantages of use of high tunnels for crop production can be as under:

1. Increased temperature and cold protection are the most significant benefits to using high tunnels. Daytime temperatures are often 30-50 degrees warmer on a sunny day, rapidly increasing plant growth, and cold hardy crops can be harvested throughout the winter. Regardless of the region, high tunnels help growers produce crops outside the normal season.

2. High tunnels provide protection from extreme weather such as high winds, heavy rain, hail, snow and drought. They also provide shade, protecting crops from heat and sunscald. Working in high tunnels allows the farmer to work/harvest in inclement weather. Ventilation can be a challenge with high tunnels. Consider installing sides that can easily be rolled up during warm sunny days to prevent high tunnel from overheating.
3. High tunnels can help reduce pest, disease and weeds in high tunnels. Row covers/low tunnel used in the high tunnel can also help reduce the need for pesticides and/or labour, producing a higher quality crop by creating or maintaining favourable temperature during peak winter or low temperature spell. Insect screens installed along the ventilators/sidewalls and doorways can exclude larger insects/dogs/cats from entering the tunnel when sidewalls are up. Some pests (mites, thrips and aphids) and diseases (powdery mildew) thrive in high tunnels and can easily get out of control. It's important to monitor crops and implement crop rotations to reduce the risk or use preventive measures such as sanitation.
4. A longer growing season, faster growth and higher productivity due to high tunnel effects lead to higher quality, and higher marketable yield. There may also be a better price premium for out-of-season crops and a potential for year-round crop/income.
5. Some plants are self-fertile, however, pollinators can increase yields significantly. Pollinators will be able to enter the structure when fully opened, but they will not be able to enter if insect screens are being used. Planting a pollinator habitat, such as flowering cover crops, around the structure will attract them. In large high tunnels bee hives can be kept for pollination
6. Since crops will not be exposed to rainfall, irrigation is essential. Arrangements of appropriate irrigation system is to be made keeping in view the climatic conditions.
7. Growing in a high tunnel requires a different set of skills, knowledge, management, tools, and equipment than in field production. Fertility, irrigation (drip), pest management, cultivars, and spacing (including trellising) are often different in tunnels from techniques used in the field.
8. Managing a high tunnel also requires intensive and vigilant attention from structural maintenance to manual labour during the growing season, and time in the "off-season" if you plan to grow year-round. Summer months are difficult to use high tunnels. Inside provision of shade net can be of use. It would be better to follow solarisation during peak summer months to keep soil borne pathogen at bay

Insect Proof Net House

Insect proof net house is a closed crop production system with 6 -8 feet height that excludes insect pests via physical barrier of synthetic insect proof 40 mesh, 20 % shade. Double layer net houses are also available which lowers the temperature in summer months. Net houses are increasingly becoming popular because of their performance and low cost of construction. Open top insect proof net houses with more than eight feet height have been found suitable in protecting crops from insects and insect vectors. Most of the agricultural insects are incapable of flying more than eight feet height.

Generally insect proof net house for growing crops is a framed structure made of materials such as GI pipes, angle iron, wood bamboo or other suitable local material. It is covered generally with plastics net (nets are made of 100% polyethylene thread with specialised UV treatment) having different shade percentages. It provides partially controlled atmosphere and environment by reducing light intensity and effective heat during day time to crops grown under it. In these structures round the year seasonal and off-season cultivation is possible. It creates favourable micro climate for the plant where the insect pest incidence is higher by mechanically avoiding their approach to crops. These nets are available in 40-50 mesh size.

UV blocking net are also available which reduce insect like white flies (*Bemisia tabaci*- vector for virus diseases) invasion by twofold to threefold than a standard 50 mesh net. Moreover, covering greenhouses with UV blocking insect proof nets can reduce both attraction and invasion of aphids (*Aphis gossypii*) and thrips (*Ceratothripoides claratus*) than standard nets of the same density

In general, sucking pests are most sensitive to radiation in the UV (330–350 nm) and in the green-yellow (520–550 nm). Thus, covering material for optically manipulating pests should contain selective additives that sucking pest can detect.

Depending on the screen material and its properties, screen houses are used to achieve various objectives, such as: exclusion of virus transmitting insects and birds and consequently reduced pesticide requirements, vulnerability of hail and wind damage, extension of the growing period and delay of fruit ripening; reduction of radiative heat loss and cooling at night and shading from supra-optimal radiation. Aluminised screen are most effective in reducing frost damage. The increasing popularity of net houses derives mainly from their fulfilling the above purposes at much lower cost of conventional greenhouses.

Virus free Kinnow saplings in insect proof net house in Haryana

Shade Net Houses

These are framed or inflated structures covered with nets to protect the plant and fruits from insects, birds, animals and extreme weather conditions. Depending upon the use, these nets are available in different mesh size and shade factors between 15% to 90%. Presently shade nets are available in white, black, red, blue, yellow and green having response on crop/plant yield attributes. It provides partially controlled environment by reducing the light & temperature favourable for crop growth.

Shade net house are considered as one of the major technologies to provide development of healthy grafts/ seedlings, potted foliage and flower plants, medicinal and spices crops and hardening for various horticultural crops irrespective of climatic conditions. Shade nets are available with different shade percent, 20, 40, 70 etc and colours. Shade net houses are made keeping in view the requirements of crop to be raised generally with 6 to 8 feet height.

Shade net is becoming a commonly used method for multiplication of horticulture nurseries and crop cultivation. Shading nets can control plant growth by reducing the light intensity and modifying other micro-environmental conditions such as air and soil temperature, air humidity, carbon dioxide (CO2) concentration, air velocity and ventilation rate. It can decrease water loss (evapo-transpiration) from plants and soil and increase CO2 assimilation. The microclimate modifications as a result of especially the reduction in solar radiation, wind speed and temperature, and the increase in absolute humidity, decrease the evaporative demand, i.e., the dryness of the atmosphere. The lower evaporative demand under shade allows plants to increase stomatal conductance and hence CO2 assimilation compared to crops in the open.

Photo-selective screens induce modifications in vegetative and fruit attributes. While comparing different coloured nets with same shading factor, researchers revealed that, red and yellow nets kindle the vegetative growth rate and vitality of foliage and cut flower crops, while the blue nets caused dwarfing, and the grey nets enhanced branching. Furthermore, pearl nets have the greatest light-scattering capability in the visible range and also absorb light in the ultra-violet (UV) range, thus found to best increase fruit size and yield in fruit tree crops, as well as postharvest quality of fresh produce.

Shadenets are available in different shade percentages or shade factor i.e 15%, 35%, 40%, 50% 75% and 90% (for example 35% shade factor means - the net will cut 35% of light intensity and would allow only 65% of light intensity to pass through the net). Each plant has its individual requirements for sunlight and shade under which it flourishes at its best. To create optimum climatic conditions, selection of the correct percentage of shade factor plays an important role to enhance plant's productivity to its optimum.

Advantages of Shade Net

- It helps in cultivation of flower plants, foliage plants, medicinal and aromatic plants, vegetables and spices.
- Shade net is used for raising nurseries of fruits, flowers, herbs and vegetables and ornamental plants.
- Shade net are helpful to enhance crop yield during summer season and facilitate production of off season vegetables
- Provides protection against pest attack.
- Protects crops from natural weather disturbances such as wind, rain, hail, frost, snow, bird and insects.
- Used in production of graft saplings and reducing its mortality during hot summer days.
- Used for hardening tissue culture plants.
- Helps in quality drying of various agro products particularly seed spices.
- Helps in creating favourable micro environment for production of vermi compost.

Uses

- Nursery propagation, floriculture, indoor plants, vegetables, tea, coffee, spices production.
- Hardening of plants, tissue culture or imported.
- Used as cattle shed, to cover fish pond, side covers in poultry farming during winter.

- Drying of various agro produce.
- Acts as a wind breaker.
- Useful greenhouse, terrace gardening and fencing
- Used as safety nets for building repairs and constructions
- Helpful in vermi- compost unit

Types of Shade Net

Presently shade nets are available in different colours i.e. white, black, red, blue, yellow and green and in combinations. Shade nets are used as

Green x Black - cut off un-wanted U.V rays and gives aesthetic look. Used in grape for providing shade and helps in drying.

Black x Black - it absorbs and radiates heat inside the shade net house. Useful in nursery raising.

White x Black - diffuses the light inside the shade net house. Mainly used for flowers such as Gerbera and Anthurium

Green x Green - enhance the process of photosynthesis in plants resulting better foliage in ornamental plants.

Tomato and cauliflower nursery under shade net – Sultanpur district, UP

Naturally Ventilated Poly houses

As the name indicate these are polyhouses with desired level of ventilation both at the top as well as side walls depending upon the geographical location of the structures. These are naturally ventilated green/poly houses with single or more span and total central height of 5 meters, side ventilation of 3 meters with roll-able poly-cover with or without roof ventilation and double/triple door entry. Such poly houses are found suitable for round the year production of crops depending on the geographical area. There can/should be multi-span naturally ventilated poly house with a central height of 6.5 meters, gutter height of 4.25

meters, side ventilation of 1.5 meters and resistant to wind storm. Generally multi-span poly houses are considered more suitable and economical for raising crops. These polyhouse do not have any environmental control system except for the provision of adequate ventilation and fogger system to prevent basically the damage from weather aberrations and other natural agents. Low pressure drip irrigation as well as fertigation systems are integral part of these polyhouses. In most of the states of India such polyhouses are popular and erected mainly by progressive farmers with or without subsidies offered by state as well as central governments. Naturally ventilated poly houses have been found suitable for cultivation of large number of vegetables, flowers and strawberries, papaya, pomegranates besides raising their nurseries in many parts of the country. Most of foliage plants are grown under these structures. For multiplication of cut flowers for domestic and abroad these polyhouses works out to be cheaper and affordable. The claddings of these polyhouses are provided keeping in view the crops to be grown and climatic conditions of the area. The structure is made strong enough to face strong winds and inclement unexpected weather.

Gerbera under poly house of Mr. Moinuddin field.- Barbanki

Technical Specifications of Naturally Ventilated Poly House (Tubular structure)

A. General Specification

SL.No	ITEM	Specification
1.	Type	Multi span (gutter connected), top ventilated, tubular structure (round / rectangle), with aerodynamic corridor/balcony along the gutter line
2.	Size	Length: Multiples of 8mx4m (Length is the side along the gable or side along the truss line). Width: Multiples of 4m (width is the side along the gutter or side along the purlin line).Gutter length to be kept within 52m as far as possible.Minimum size would be 500 sq m.

Contd.

3.	Grid size and shape	Grid: 4m X 8m; preferably rectangular shape
5.	Height	Gutter level: 4.0m minimum from ground level at any point.Ridge level (central height): 6.5m - 7.0 m.
6.	Orientation	North to South (Direction of Gutter run) is preferable. Or as per the local condition depending on available land, wind direction, land slope etc.
7.	Structural component	Hot dipped, Galvanized Iron, IS 3601:2006/IS1239:2004, GI pipe wall thickness not less than 2mm for all the structural components. Galvanization of structural members should not be less than 300GSM (grams per square meter). Structural member should be preferably connected with clamps, nuts & bolts and fasteners properly.

S.N.	Component	OD (inch)	OD (mm)	Minimum wall thickness (mm)	Weight per meter length(Kg)
A	Column Foundation pipe	2½	76	3.0	5.36
	Corridor Foundation pipe	2	60	3.0	4.20
B	Column (pillar)	2½	76	2	3.75
C	Arch Purlin	1½	48	2	2.30
D	Gutter purlin	1¼	42	2	2.1
E	Boundary or perimeter purlin	1¼	42	2	2.10
F	Balcony/corridor/hockey column	2	60	2	2.85
G	Corridor horizontals	2	60	2	2.85
H	Top and bottom arch of truss	1¼	42	2	2.10
I	vertical and inclined members of truss	1	33	2	1.60
J	Bottom chord of the truss	1½	48	2	2.30
K	Arch brazings, V brazing, X brazing	1	33	2	1.60
L	Crop support purlin	1¼	42	2	2.10
M	Rolling pipe	¾	27	2	1.30
N	Top vent brazing/sturt	1	33	2	1.60
O	Curtain shaft & handle	¾	27	2	1.30
M	Curtain guide	½	22	2	1.00
8.	Top or ridge ventilation	Top (roof vent): Top vent of 100 to 120cm of vertical gap, 120 to 150cm slanting length. Covered with UV stabilized insect proof net (40mesh) / anti hail net /monofilament /tape shadenet as per crop demand. Minimum overlap of top arch and bottom (small) arch should be 60cm to avoid direct rain entering into the greenhouse from the vent. Top vent purlin brazing to be provided with ¾ inch GI pipe			
9.	Side ventilation	3.25m-3.75m down from gutter level. Covered with 40 or 50 mesh UV stabilized insect proof net or 35%/50% monofilament / tape shadenet as per crop demand. Over to that rollable polyethylene flap has to be provided (200 micron UV stabilized) with cranking arrangements.			
10.	Balcony/corridor	60mm OD/2mm thick GI pipes (2m+2m along the gutter line)			
11.	Withstand to Wind Velocity	The structural design need to be sound enough to withstand wind speed of 130 Km			

B. Specification of Gi- Material for Fabrication

1	Foundation	GI- Pipe = 2½" (OD 76 mm) 3.0 mm thickness or 60mm OD & 3mm wall thickness, 1.2-1.5 meter length, crimping / telescopic with nut & bolt. Foundation pipe must be provided with 6 inch length 12mm TMT anchor rods at 3 locations. All gutters must have a unidirectional slope of 1 to 2%. The column foundation should be of at least 45 cm diameter x 90 cm depth and filled with cement concrete Ratio of 1:2:4.
2	Columns	GI- Pipe = 2 ½ (OD 76 mm), not less than 2.0mm wall thickness (minimum 4 meter length, no joints); on the gable side, provide 6.5m columns between two gutter columns.
3	Arch Purlin	GI- Pipe = 1 ½" (OD 48mm), not less than 2.0mm wall thickness (in 4 meter length), Arch Purlin is a member which inter connects arches.
4	Gutter Purlin	GI- Pipe = 1¼" (OD 42mm), not less than 2.0mm wall thickness (in 4 meter length), gutter Purlin is a member which inter connects base of arches and also supports the gutter. Gutter purlin is not required if gutter sheet of above 1.6mm thickness is used.
5	Purlin at gutter level (end bays)	GI- Pipe = 1 ½" (OD 48mm), not less than 2.0 mm wall thickness (in 4 meter length).
6	Boundary purlin/ perimeter purlin	GI- Pipe of 1¼ (OD 42mm), not less than 2 0mm wall thickness (in 4 meter length), connecting all the columns in the periphery or boundary.
7	Trusses	GI- Pipe of 1¼" (42 mm OD), thickness 2.0 mm for top and bottom arch. Bottom horizontal of the truss must be 1½ (48mm OD) and all Vertical and inclined members should be of 1" (33mmOD) and 2.0 mm thick.
8	Rolling Pipe	GI- Pipe of ¾" (OD 27mm), thickness not less than 2 .0mm (in 6 meter length) for rolling polyethylene flaps for side vent, provided with cranking arrangements. A GI pipe (22mm) guide need to be provided at the corner and other suitable places.
9	Note	Welded pipes should not be used for structure erection (except horizontal bottom chord of truss of 8m length, rolling pipe and longer column on the gable side)

C. Specifications of Gi- Material for Reinforcements

1	Arch (truss) Brazing	GI- Pipe = 1 inch (OD 33mm), thickness 2 .0mm (6 meter). Connects the arches (at vertical chord joining point) to the top of 2nd pillar on gutter line. Provided to both the gable end bays.
2	Column Brazing / X Pipes	GI- Pipe = 1 inch (OD 33mm), thickness 2 .0mm for corners & last but one span all along the gable line.
3	V – Brazing	GI- Pipe = 1 inch (OD 33mm), wall thickness 2 .0mm). Interconnecting columns to bottom chord of truss.
4	Top vent purlin brazing	GI- Pipe = ¾" (OD 27 mm), thickness 2.0 mm (1.2 to 1.5m). Connects the bottom and top arch in the vent.
5	Crop support purline	GI- Pipe = 1½ inch (OD 42mm), thickness 2 .0mm (4 meter) Connect gable side pillars below the bottom chord for running GI wire grid for crop support.

Contd.

6	Clamps and Nut Bolts, brackets, cleats, fasteners	Different type of clamps like 76/60/48/42/33 mm OD full, 76/60/48/42/33 mm OD half are used which should be made from min. 40-45 mm wide and 2.0 mm thick Galvanised Plain coil having minimum 200 GSM Galvanization. Brackets and cleats must be 100mm X 100mm X 6mm. Nut Bolt: 3/8size, length 1.5" & 3½". Galvanized material should be used.
D.	**Specification of Fixing/ Clading Materials/Other**	
1.	Gutters	The gutters are made of GI sheet of 1.6mm thick, trapezoidal shape, 50mm perimeter or more, if supported on columns. Optionally 0.5-0.6mm thick aluminum coated (Galvalume) jointless (single length) sheet of 60mm perimeter supported with two 42mm OD X 2mm thick GI purlin fitted with plated SDS screw. Option of plastic gutter (single piece), Vergin, UV stabilized, 1.4mm thick, 60cm wide supported by Gutter purlins is also provided. Unidirectional slope for all the gutters, minimum uniform slope of 1 to 2% is to be provided in civil structural work/design. Gutter at the lower ends must extend at least 50 cm to avoid rain water falling on the pillar foundation. Total gutter length preferably not more than 52m. An option of external gutters/end gutters on two sides (parallel to gutter line) to maximize rain harvest and to prevent rain water entry through corridor/pavilion can also be provided. Lower end of the gutter to be connected with GI funnel to minimum 110mm PVC (4.0kg/cm2) vertical pipe (4.5m) leading to ground level drainage (nala).
2.	Polyethylene film (Plastic)	For the roof, side walls and curtain walls: Pre requisite propertiesUV stabilized, 200 micron thick, 3/5 layered, anti drip/ mist, anti dust, diffused 50 to 70 %, having minimum 85% level of light transmittance.Optional property • Anti sulphur for the crops where sulphur consumption is high. • For ex - rose cultivation. • IR Reflective cooling –for high temperature area.While installing the multilayer film, first ensure that respective layers are facing the right direction as shown on film (e.g. inside out)The logo, brand name and batch no with code number must be printed at regular interval on poly films that should not get washed easily. Certificates to be provided: 1) Test report from Central Institute of Plastics Engineering Technology (CIPET). 2) Technical Data Sheet
3.	Side Apron / compound sheet	200 micron UV stabilized poly film or UV stabilized HDPE woven fabric, 150 gsm thick for a height of 75 cm above ground level and 30 cm buried below ground and 20 cm horizontally.
4.	Aluminum Profile	The polyethylene coverings and the nets are fixed to the structure by a locking profile made of extruded aluminum profile. Made of alloy aluminum (approximately 125g/running meter). Should be suitable for double spring locking.
5.	Zig-zag Spring	Made of minimum of 2.3 mm-thick steel that is coated with plastic & it will lock & hold poly film and nets in aluminum profile.

Contd.

6	Fasteners (Screws to fix aluminium profile)	Galvanized Self Drilling Screw (SDS) with washers as per requirement and with the best quality plating to have good anti-corrosiveness. SDS have to be fixed on profile for every 50cm distance.
7	Side & top ventilation Net	40 mesh and white in colour on all four sides and top vent of green house (UV- Stabilized). Monofilament /tape shadenet also may be used for better air circulation as demanded by the crop. Anti hail net/ Monofilament /tape Shadenet can also be used to the top vent for better air circulation depending on the crop requirement.
8	Shade Net	Shade net vary from 35% to 50% based on crop (UV- Stabilized) which is made retractable with proper hooks & UV stabilized PP ropes. Shadenet is supported with 12 gauge GI wire. Shadenet is spread on a 4m width bay at gutter level. Shorter Retractable arrangements with pulleys for easy retraction and spread.
9	Crop support for vegetable cultivation	Trellising system to be provided with 10, 12 gauge GI wires at required spacing & height as per crop requirement.
10.	Rollable polyethlene flap for side wall	3.25 to3.75m down from gutter level with cranking arrangements and handle for all the sides (flap control pipe). ¾" plastic side curtain clamps are used to hold plastic & GI-Pipe at every 30cm. Flap curtain guide of ½ " (22mm OD) pipe to be provided at regular interval for better rolling arrangements.
11	Door (using square pipe).	Double Door Entry System, Sliding Door, Polycarbonate sheet (both side U V stabilized, 4mm thick) on top and aluminium sheet at bottom (4mm thick) fixed to suitable frame. Minimum 2m height, 1.5m width, with locking arrangement.
12	Ante room (Entrance room)	Size should be at least 4 m length x 2m width, 2m height, Ante room door situated diagonally opposite to poly house door. Flooring: 50mm PCC flooring over 75 mm thick sub base. Provision for foot wash basin (2 feet x 3 feet x 0.5 feet depth) near outer door inside entry room.
13	Drip System installation	16mm Inline drip lateral, class-2, Discharge: 1 to 2 lph, Emitter spacing 20-40cm, PVC mains/sub-mains, control valves, flush valves, proper sectioning based on water availability, disc filter and ventury system/fertilizer injector with pressure gauge. 5600m inline drip lateral pipe need to be provided for 1 acre area.
15	Fogging system (optional) Used only for climate maintenance, not for irrigation	To be provided depending on crop requirement. It consists of 4-way antileak fogger of 10-28LPH flow rate, to achieve particle size of 80-100micron, 16mm LLDPE lateral pipe, 6kg/cm2 PVC pipe, valves, pump, starter, control panel, MCB, relay, complete to apply 3mm/hr.
16	Rain water harvesting – Desirable	Provisions must be made for collecting the rain water from the roof top. HDPE Geo Membrane sheet with 300 micronand above with BIS or equivalent certificate to be used for lining the pond of appropriate size to hold 25% of the normal annual rainfall of the region
17	Instruments	EC meter, pH meter, Hand held digital RH and temperature measuring instruments.

Source: National Horticulture Board

Rain shelters

Rain shelter, a low cost naturally ventilated poly house is found ideal for off-season vegetable cultivation in Kerala, Andaman and Nicobar Islands, north east region and other high rainfall areas of the country. Off-season cultivation of vegetables by utilizing low cost protected structures like rain shelter can be considered as a profitable enterprise besides protecting crop nurseries from high rain during monsoon months. Such structures are boon for horticulture production in high rainfall areas of the country. The rain shelters design is developed on farm and local material is used for support of cladding material which is UV grade plastics by and large. Other claddings are also used. Rain shelters are impervious to rains exposed to the free entry of insect –pests. In certain structures insect prof netting is used but it does increase temperature and humidity inside the structure. Rain shelter houses are roofed with plastic film and other waterproof materials to shelter crops from rain. The houses are effective in reducing crop damage caused by heavy rains in promoting crop growth, and in achieving stable production of high quality vegetables. Thus, the use of rain shelter houses are highly effective measures in vegetable production. Particularly in heavy rain areas, its effectiveness is quite high. However, the cultivation under rain shelter houses involves several problems. It induces higher cost of production as compared with the open field cultivation, and the houses are liable to be damaged by strong wind. Furthermore, advanced techniques in watering and fertilizer application are required for the success of the cultivation under rain shelter houses.

Rain shelters for crop production

Innovative Structures or greenhouses

There can be several different innovative structures suiting to local environment/ climate. Two of such structures i.e soil trench /underground poly house and Ladakhi polyhouse ideally suited for vegetable cultivation in cold desert of Ladakh, have made cultivation of all sort of vegetables possible which was considered in 1980s as impossible due to around six months' sub-zero temperature prevailing atmosphere there. This contribution of DRDO laboratory named Defence Institute of High Altitude Research (DIHAR), Leh is unique globally.

Soil trench or Underground or Sun-pit Greenhouse: Since cropping period is small in Ladakh and other high altitudes, it was difficult to raise vegetables requiring transplanting, such as cauliflower, cabbage, knol khol, broccoli, kale, tomato, capsicum, chili, brinjal, true potato seeds etc. Defence Research and Development Organization (DRDO) has standardized soil trench cultivation method of raising early nursery of vegetables. Seedlings are raised in polythene covered trenches of appropriate sizes, generally 15x5x3 or 30x10x3 feet during March and April and transplanted as soon as cropping season starts. . Trench is covered with transparent UV-stabilised. 200 m thick polythene during daytime to harvest maximum solar energy and extra cover of black polythene at night time to check the heat loss

This is a unique, innovative, very simple, cheap, and useful underground greenhouse structure for the hilly region and thus has unlimited potential in the region. As mentioned earlier this may be of any convenient dimension. In this pit type of structure, wooden poles are used to hold UV-stabilized polyethylene film. The polyethylene is also covered by an additional or woollen or cotton sheet polyethylene film during night to reduce the heat loss during extreme winter. The damage of blowing off the polyethylene film by strong winds is minimized by putting stones along the sides. Cultural practices and other operations are done by removing the polyethylene sheet from the top of the trench. The structure does not require much skill in its construction and management. Its cost is lowest among all other greenhouses, and being an underground structure, heat loss is minimal and temperature retention is high (Singh and Dhaulakhandi 1998) and thus yields good crop. Strong winds do not affect polyethylene cover much and hence it is long lasting. This structure is therefore being recommended as the most suitable greenhouse for the cold desert of India where precipitation is very very low or not there (Singh et al. 1998, 2000). In Ladakh these trenches during summer are used for cultivation of cucurbits and Solanaceous vegetable crops. Trench can be used to raise off season potato crop (March- May/June) besides raising leafy vegetables during severe winter.

Ladakhi Poly house: Another structure, Ladakhi poly house made of unbaked bricks on three sides and polythene on southern sidemeasuring 12x4 x2 m size is made using local poplar and willow wood. It is covered with ultraviolet stabilized white polyethylene film of 200-micron thickness or other films like rigidex. Crop inside polyhouse is covered over night with 150-gauge black polythene. During frozen winters this structure like soil trench produces leafy vegetables (palak, mustard, fenugreek, celery, parsley etc). Ladakhis used to laugh on the idea of producing vegetables in long sub-zero atmosphere of Ladakh. Now they themselves are practicing it to meet their kitchen requirement and that of local defence forces.

This is one of the innovative and low-cost greenhouses of the Ladakh region. It is similar to walk in tunnel but having surrounding mud brick wall in place of polyethylene sheets which not only cuts down the installation cost but also reduces the adverse effects of strong winds and also increases temperature retention and ultimately increases net profit. This is generally the lean type and has three sides made of mud bricks. The back wall is 7 ft in height, while the front has no wall. The length is 32 ft with a width of 16 ft. The dimensions of the structure can be adjusted to the space available. The polyethylene is supported on wooden poles and sidewalls. The two sidewalls are descending toward the front. Local administration /Jammu and Kashmir State Department of Agriculture/ Horticulture provides cash subsidy besides the 32 × 16-ft polyethylene sheet. This has led to vegetable production in winter frozen months.

Double-Wall Polyench

This structure is designed by the Forest Research Laboratory (FRL) now known as Defence Institute of High Altitude Research (DIHAR), Leh, to harness the soil and solar heat for growing the vegetables especially during winter months. The polyench refers to a polyhouse erected over a trench where back walls and sidewalls are made up of mud bricks. The sidewalls are provided with double walls fill with insulating material available locally, i.e., sawdust for better retention of heat inside the polyench. The inner walls are painted black to absorb more solar radiation during morning hours. The structure is also like gable uneven span with modified roof having a polythene sheet toward the sun-facing side and grass thatch support over a wooden frame opposite to the sun- facing side with provision of ventilators in the roof. Since locally available material has been used in this greenhouse, the cost is reduced drastically compared to a normal greenhouse with better efficiency of temperature retention than the trench and polyhouse. Since locally available material has been used in this greenhouse, the cost is reduced drastically compared to a normal greenhouse with better efficiency of temperature retention than the trench and polyhouse. Considering the wear and tear that occurred due to high wind velocity in the hilly regions, the polythene sheet has been replaced with FRP (fibre reinforced plastic) sheets to increase the life span of the structure.

FRP (Fibre-reinforce Polyester) Greenhouse

DIHAR has designed, evaluated, standardised and disseminated FRP greenhouse technology to to Leh and Kargil farmers and interested. Defence Forces. FRP greenhouses of size 100 x 30X 10 ft have been found suitable for the region.

Solar Greenhouses

Though all greenhouses are designed to collect and use solar energy, solar greenhouses are specifically designed to trap and store thermal energy from the sun. This stored energy can be used later at night or when it is cloudy outside. Solar greenhouses can be in the form of gabled or arched structures, cold frames or even subterranean pits. Large scale solar greenhouses are usually free standing, while backyard solar greenhouses are attached.

Solar greenhouses may be of two types: active or passive.

Passive Solar Greenhouses

Passive solar greenhouses work by absorbing solar radiation through glazing, converting it into heat and limiting its leakage back to the outside. Much of the sun's energy is absorbed by plants which then radiate it back into their surroundings. This radiated heat is absorbed by surfaces such as soil, water, rock and other inexpensive materials. When temperature falls, these materials passively allow heat flow to the greenhouse environment to keep it suitably warm. No external energy is needed to maintain an ambient temperature in this type of greenhouse except in extremely cold conditions when electric or gas heating may be needed to supplement solar heating. Most of the solar greenhouses discussed above are passive greenhouses.

Active Solar Greenhouses

Active solar greenhouses also collect, store and deliver energy like the passive ones except that they need supplemental energy to direct the thermal energy from storage areas to the greenhouse environment. Of the two types of solar greenhouses, the passive ones are less costly.

Poly house with Pad and Fan System: These poly houses unlike naturally ventilated poly house has fan and pad system to regulate temperature and humidity. They have generally height of 4.5 meters with common side and top ventilation. These structures are clubbed with exhaust fans and cellulose cooling pads of 1.8 m in height and 150mm thickness. The cost of construction, running and maintenance is high. Lack of assured power supply becomes a limiting factor for such structures in our country.

Retractable roof poly houses are also becoming popular in certain areas where day temperature is high and night temperature is lower resulting in natural favourable conditions for the plants to grow and yield better at lower cost but initial cost of structure is high.

Best Option 2019

Best options on date for scaling out greenhouse/protected cultivation among farmers are:

1. Net house + plastic mulch + fertigation for large number of crops like tomato, cucumber, capsicum; chrysanthemum, lilies, roses, gerbera, anthurium, foliage plants; stevia, basil, papaya, strawberry, pomegranate and others.
2. Plastic tunnels both low as well as walk-in tunnels are most suitable for temperate climate with modifications according to climate of the place and raising early cucurbits in northern plains. Rain shelters have been found successful in high rainfall areas like Kerala.
3. Climate controlled greenhouses are needed for raising plant nurseries to meet huge seedlings requirement of vegetable and flower growing clusters both under open field cultivation and cover besides ornamental/ foliage plants multiplication.
4. Inputs and guidance for these from reputed private and public sectors are must.

26

Greenhouse Heating Systems

There are several systems for greenhouse heating based on heat source and purpose or crops to be produced in greenhouse.

Electric heaters

Unit space heaters, either floor mounted or supported, are normally dependent on electricity supply supported by stand-by generators. Abroad they are fuelled with natural or bottled gas or fuel oil and use fans for heat distribution. This system requires a relatively moderate capital investment, is easy to install, and provides for easy expansion of facilities. If unit air heaters are used, they should be spaced and directed to blanket the entire area with heated air.

Hot Water Systems

Hot water systems utilize metal pipes that can be perimeter, under benches, or overhead fan forced unit heaters can be used. This system requires a boiler, valves, and other necessary controls. Hot water system is simpler to install and normally requires less maintenance than a steam system. Hot water systems are mainly used in smaller greenhouse.

Steam Heating System

A steam heating system needs a boiler, valves, traps and other controls depending upon the size and type of boiler used. Steam provides rapid heating and cooling of the steam lines and usually less pipe is needed. Lines may be smooth or finned, and about 1/3 of the heat should be overhead and about 2/3 along the side walls. Lines can also be arranged under benches or with overhead fan-forced unit heaters. A steam system also allows the use of steam for soil pasteurization. A steam system requires a high initial investment; however, it has a long life expectancy. Steam heating systems are most often used in large ranges as steam can be transported long distances efficiently. Geothermal heaters also come under this category and common in Norway and Ice land and wherever geothermal energy is available.

Unit Radiant Heaters

Unit radiant heaters using gas of the "Salamander" grove heater (return stack) type and which use fuel oil may be used for frost protection. These are not recommended for routine greenhouse heating because they are not thermostatically controlled making temperature control difficult. Also, because they are not vented to the outside of the greenhouse, combustion gases will damage sensitive plant material. They can be used for emergency protection against cold temperatures. These systems are uncommon.

Solar Heating

On sunny days' temperature/heat inside the greenhouse increases which is to be trapped by several means to be slowly released during light when temperature goes down to undesirable level. Plastic water bottles or containers of appropriate size are placed inside the greenhouse to absorb heat during the day and release during night. Ladakhi or chineese greenhouses have three sides of greenhouse mud walls painted black which absorb and release heat during day and night, respectively.

Where ever geothermal heat/steam is available, it is passed through the greenhouse evenly mostly during dark period to maintain greenhouse temperature.

According to climate heating of greenhouses become necessary to maintain temperature inside the greenhouse round the clock above 18 degree Celsius and desirable level of relative humidity. To achieve these conditions, a greenhouse heating system is in most cases is required. The heating of a greenhouse is traditionally done by using a central gas boiler which heats water that is distributed into the greenhouse. This radiant heating method works on the same principal as the average central heating installations in homes. The temperature and flow of the heated water throughout the greenhouse can be accurately controlled with automated pump- and valve systems.

In some cases, heating the greenhouse is done by installing air heaters in the growing areas of the greenhouse. These hot air heaters run on either electric or natural gas, petroleum and diesel. The most common heating system in commercial greenhouses is the multi-purpose tube rail. Especially in vegetable crops, the tube rail heating system has been widely applied as it has a significant logistical advantage. Another common hot-water circuit in the vegetable production is the grow tube. The grow-tubes are situated at the fruits throughout the greenhouse which enables the grower to control the ripening process. In severely cold climates, the tube-rail system and the growing tube are supplemented with a snow-tube circuit. This to melt off the fallen snow from the deck of the greenhouse, without having to open the climate screens and to disorder the temperature at crop-level. In India heating of greenhouses has not become popular yet.

27

Greenhouse Drying

About 80% of the world population live in developing countries and are dependent on agriculture. Agricultural products, just after harvesting, are dried to the safe moisture level. Open solar drying for drying agricultural products is being practiced since ages throughout the world. Due to its several drawbacks as danger of (dust, rain, ultra-violet rays, birds, animals etc.), advance technique, i.e. greenhouse drying, is being adopted for drying crops to reduce the drying time and increase the quality of the food products.

Drying involves a heat and mass transfer phenomenon in which heat energy supplied to the product surface is utilized in two ways: (i) to increase the product surface temperature in the form of sensible heat and (ii) to vaporize the moisture present in product through the provision of the latent heat of vaporization. The removal of moisture from the interior of the product takes place due to induced vapour pressure difference between the product and surrounding medium. The moisture from the interior diffuses to the product surface to replenish the evaporated surrounding moisture.

The working principle of greenhouse drying is in which the product is placed in trays receiving the solar radiations through the plastic cover and moisture is removed by natural convection or forced convection.

Greenhouse drying are being adopted to reduce the losses and to increase the quality of the dried products. Solar energy is the most promising power source of energy for drying of agricultural products. Most of the farmers adopt open sun drying because it is abundant and inexpensive. But the open sun dried products are also subjected to remarkable losses due to environment. Losses of fruits and vegetables are estimated to be 30%-40% during drying under open sunlight. The post-harvest losses can be reduced by adopting advanced means of drying, i.e. greenhouse drying. The greenhouse dryer can be operated in natural and forced mode as required. Studies reveal that the greenhouse dried products are of superior quality and better colour retention as compared to open sun drying.

Now a day, the demand for dried agricultural products, food grains, vegetables, fruits, herbs, spices and so on has increased. Traditionally, products are dried in the open sun and are cheap, but the quality of these products is deteriorated by ultraviolet rays, dust, insects, animals, microorganisms, etc. So, the open sun dried products are not meeting the international standards. Off- season cultivation of agricultural products in controlled environment is also increasing. Therefore, the advanced means of drying, i.e., greenhouse drying are being adopted to reduce the losses and to increase the quality of the dried products. The greenhouse dryer can be operated in natural and forced mode as required. Greenhouse dried products are of superior quality and colour as compared to open sun drying. Most of foods gains, fruits, vegetables, herbs, spices, flowers, fish, mushrooms can be sun dried under different solar driers.

Several crop solar dryers exist. Some of them are listed below.

- The natural convection and forced convection dryers.
- Direct solar dryers, whereby products are exposed directly to solar radiation and indirect type solar dryers by which the products are not exposed directly to solar radiation but the air heated by the solar radiation is made to flow through them.
- Integral dryers.
- Greenhouse dryers.
- Cabinet dryers.
- Tunnel dryers
- Mixed mode dryer in which the products are exposed directly to solar radiation and also to hot air flowing through them.

Greenhouse dryers

Hybrid sustainable greenhouse dryer can be beneficial in rural regions to reduce the spoilage and improve the quality of the dried products so that the farmers get the significant agricultural return on their efforts. It has been observed that greenhouse drying technology significantly improves the quality of the products and reduces the drying period.

Garlic drying in greenhouse

Greenhouses of different shapes have been used for drying of various products under different environmental conditions. Photovoltaic/thermal greenhouse dryer are also used. The greenhouse dryers were observed to be 2-5 times more efficient than open sun drying mode.

Polycarbonate cladded greenhouse for drying chillies

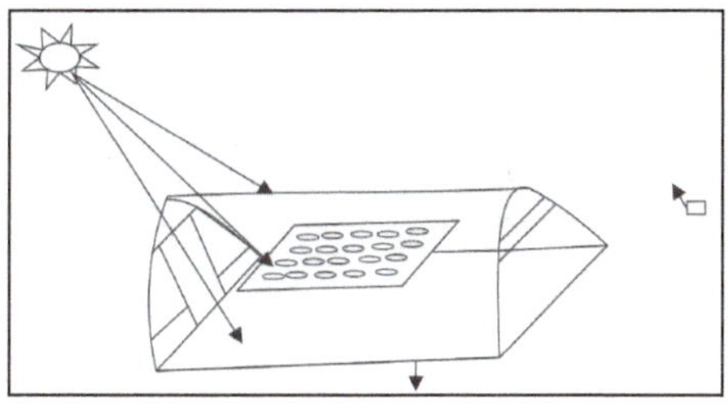

Solar tunnel dryer drawing

Greenhouse production and drying of Moringa leaves, Chennai

Greenhouse dryers are constructed with a principle that they have a rigid load-bearing frame which is placed at certain spans and not deformed under the loads acting on them and a transparent cover material placed on them which allows short wave solar radiation to enter and is partially opaque to the long wave radiation leading to a greenhouse effect. Design of greenhouse should depend upon sound scientific principles that facilitates controlled environment for plant growth. Orientations of greenhouses are negotiating for wind direction, latitude of location and type of temperature control and avoid dropping of shadow on the nearby greenhouses. Design of greenhouse, the shape and orientation of a greenhouse plays an important role.

Table: Moisture content details of various crops for solar drying

Crop	Crop Initial moisture content (% w.b.)	Final Moisture content (% w.b.)	Maximum allowable temperature (°C)
Paddy, raw	22-24	11	50
Paddy, parboiled	30-35	13	50
Maize	35	15	60
Wheat	20	16	45
Corn	24	14	50
Rice	24	11	30
Pulses	20-22	9-10	40-60
Oil seed	20-25	7-9	40-60
Green peas	80	5	65
Cauliflower s	80	6	65
Carrots	70	5	75
Green beans	70	5	75
Onions	80	4	55
Garlic	80	4	55
Cabbage	80	4	55
Sweet Potato	75	7	75
Potatoes	75	13	75
Chillies	80	5	65
Apple	80	24	70
Apricot	85	18	65
Grapes	80	15-20	70
Bananas	80	15	70
Guavas	80	7	65
Okra	80	20	65
Pineapple 80 10 65 26	80	10	65
Tomatoes 96 10 60			
Pineapple	96	10	60
Tomatoes	96	10	60
Brinjal	95	6	60
Peanuts	40-55	8-10	
Fig	70	20	70
Coffee	65	11	
Spinach	80	10	
Ginger	80	10	
Turmeric	80	10	
Nutmeg 80 20 65 41	80	20	65
Sorrel 80 20 65 42			
Groundnuts 40 9			
Sorrel	80	20	65
Groundnut	80	9	

Contd.

Prunes	85	15	55
Peaches	85	18	65
Mulberries	80	10	65
Yam	80	10	65

Source: International Journal Engineering Research and Technology, 2014; 3(3):1053-1066

Usefulness of greenhouse drying

1. Drying rate is faster under forced convection (greenhouse) than natural convection mode (open sun drying).
2. In greenhouse drying convective heat transfer coefficient is more than open sun drying mode.
3. Values of convective heat transfer coefficient diverse with the size and primary moisture content of a product.
4. Use of inclined reflection north wall in greenhouse solar drying can increase the total solar radiation.
5. Photovoltaic- ventilated solar greenhouse dryer gives a very high quality dried product and reduces drying time than open air sun drying.
6. For coconut drying, natural convection solar tunnel dryer is more suitable.
7. Greenhouse drying reduces moisture content in a very less time.
8. Solar greenhouse tunnel drier can be used in rural areas for drying of several agricultural produce.
9. Relative humidity under forced convection is inferior to natural convection mode.
10. Greenhouse type agricultural produce dryers result in much higher quality produce and are more effective than open sun dryers.

28

Irrigation Systems Used in Greenhouse

The following various types of irrigation methods and their systems are used in crop production in greenhouses and plant nurseries.

1. Hand Watering
2. Perimeter Watering System
3. Tube Watering System
4. Drip Irrigation System
5. Overhead Spray System
6. Ebb and Flow System.

1. Hand Watering

The most uneconomical but common and expensive system of watering. It is followed where labour is cheap and automation is not desirable due to small area/field. Hand watering in crop nurseries and potted plants is common practice. Hand watering requires considerable time and it is not pleasant work. During hand watering force of water is to be broken by using fine spay or perforations to avoid disturbance to soil/growing media and exposure of roots.

2. Perimeter Watering System

Pipes run around the perimeters of a bench with a nozzle that sprays water root medium surface below the foliage. Thus this system is suitable for production of fresh flowers.

3. Tube Watering System

This system is mostly used for watering the pots. Tubes fitted with nozzle or without nozzle are used.

4. Drip Irrigation System

This system is highly suitable for green house and open field cultivation with or without mulch. There is a saving of 50-70% of water in comparison to flooding system of watering. It provides uniform distribution not only of water but nutrients, pesticides and fungicides without wastage once used with irrigation water.

Drip irrigation system delivers water to the crop using a network of main line, sub mains, lateral line with emission point spaced at regular interval along the length side. Each dipper or emitter supplies a measured precisely controlled uniform application of water, nutrients, pesticides, fungicides and growth substances directly into the root zone of plant.

Water and nutrient enter soil from emitters moving in root zone of plants through combined forces of gravity and capillary. In this way, plants get moisture and nutrient replenished immediately ensuring plants which never suffer from water stress thus enhancing quality. This results in optimum growth and high yield.

PVC pipe line is used for distribution of water from the main source to sub-lines. Drip laterals of 12-16 mm are suitably placed in each row of plants which are directly connected to sub lines. Drippers/emitters fitted to the laterals are laid just near the plant for gradual distribution of water up to the root of plants.

There should be regular maintenance of pipes by adding chlorine or other chemicals in drift line to kill bacteria and algae. Treatment with acid is also needed to dissolve calcium carbonate. Care should be taken against rodents to avoid any damage to pipes. Monitor the pipes thoroughly for any leakage which should be done regularly. Advantages of the system with other details are given below under modern greenhouse irrigation.

5. Overhead Spray System

Green house crops are most easily and cheaply can be irrigated from over-head. In this system pipes are hanged over the plant from 60 cm to 180 cm above the plants. Nozzles are fitted in the pipes at 360° pattern. Care should be taken that the water should be of good quality and properly filtered so that these nozzles are not clogged.

6. Ebb and Flow System

This is a sub irrigation system for potted and bedding plants. Pot or flats are grown in a levelled bench made of plastic or fibre glass available in various widths (1.2-2.0 m) and length is 1 m. These benches are glued to gather to make a bench of desired length and width. These benches can move over a stand. These have channels to drain out irrigation water.

There is a water tank below the bench. This tank is covered to avoid dust or algal growth. The area of watering and size of pump should be matched. To ensure watering of pot, water remains on the bench for 10 minutes and collected in tank and reused. Fertilizer may also be applied through this system of watering. Pot with bottom hole is better than side holes.

Modern Greenhouse Irrigation

The key issues in irrigation of greenhouse crops are

1. Conserve Water
2. Improving water use efficiency
3. Sustainability of irrigated agriculture

Irrigation schedule: It means how often or when water is given, and how much irrigation water has to be given to the crop?

Irrigation methods and systems used in crop production in greenhouses are as under

1. **Surface systems:** Application of water on soil or substrate surface

   Hand watering; drip irrigation; micro spray

2. **Overhead systems:** Application of irrigation water over the top of the crop or foliage

   Sprinkler system; mist and fog system and boom system

3. **Subsurface systems:** Application of water to the base of the root zone and rely on capillary action or flooding to bring water into the root zone.

   Subsurface dip; Troughs; Flooded trays; Flooded floors

Irrigation Delivery Systems

- Pump and pressure tank may be used to increase or stabilise the hydraulic pressure
- Injectors are used to inject various nutrients into the water. Mixing or blending tanks help the dissolved materials distributed uniformly in the water.
- Filters are fitted to remove insoluble materials.
- All irrigation systems consist of tubing and/or pipe to transport the irrigation water from the source to the individual plants or a group of plants.

- Minimizing the pressure drop with the system, and especially the variation in pressure among the emitters, maximises the irrigation uniformly. It is important to design of irrigation systems properly so that all the parts (main lines, valves, pressure regulators, laterals and emitters) are correctly sized to deliver at the desired pressure and flow rate.

Drip Irrigation

Drip irrigation is the slow localized application of water, at a point or grid of points on or just below the soil surface. Water is supplied to the plats by set of plastic tubes laid on the ground or buried at a depth of 15-40 cm. Drip laterals are left in place throughout the irrigation season. Laterals are12-20 mm in diameter and either perforated or fitted with a special emitter. The discharge in the range of 1-8 litre per hour per emitter, should not exceed the soil in-filterability.

Advantages

1. Significant saving of water, nutrients and operating cost are possible if this designed and managed correctly.
2. Water use efficiency can be higher under drip irrigation because of less evaporation, continuously high soil water at near the field capacity and lower root zone salinity.
3. Reduced weed problems and non-wetted soil surface resulting in easy operation. Water distribution is highly uniform if maintenance and construction done properly.
4. Festination can easily be done with minimal wastage of fertilizers.
5. The plants and fruits remain dry, reducing the risk of diseases.
6. Ease in automation of the system

Disadvantages

1. Clogging is a serious risk if the water is not properly filtered and equipment is not properly maintained.
2. Susceptibility to clogging of dippers may lead to the more expensive system depending on the water quality.
3. Drip tubes cause extra labour after harvest and before planting.
4. Salt tends to accumulate at the soil surface and around the wetted area.

Drip irrigation systems

It is consisting of

1. Storage reservoir and/or water source
2. Control unit
3. Pumps
4. Non-return valve
5. Dilution equipment
6. Filters (sand separator, sand filter, screen filter, disc filter and others)
7. Pressure regulators
8. Water meters
9. Valves (Solenoid valve should be used if irrigation is to be automated)
10. Mainline
11. Sub-mainlines
12. Laterals and
13. Emitters

A drainage system is also required to collect the drainage solution to be reused or discarded.

A closed system, a reservoir and disinfection equipment e.g. UV, heat system or slow sand filters should be installed recycling and disinfection.

Various emitters are available in a wide range and should be selected according to the type of growing techniques.

- Punch-in-emitters and built in emitters
- Emitters having flushing features
- Non-drip emitters
- Emitters have multiple outlets

Water output in drippers

a. 16mm dripper at 2.8kg/cm^2 pressure gives 2.65 litres/hour (LPH).

b. 15mm dripper at 1 kg/cm^2 pressure gives 1 to 4 litres per hour

Filters: Depending upon the type of water, different kinds of filters can be used.

Gravel filter: Used for filtration of water obtained for open canals and reservoirs that are contaminated by organic impurities, algae etc. The filtering is done by beds of basalt or quartz.

Hydrocyclone: Used to filter well or river water that carries sand particles.

Disc filters: Used to remove fine particles suspended in water

Screen filters: Stainless steel screen of 120 mesh (0.13mm) size. This is used for second stage filtration of irrigation water.

Fertigation system

In fertigation system an automatic mixing and dispensing unit is installed which consists of three systems pump and a supplying device. The fertilizers are dissolved separately in tanks and are mixed in a given ratio and supplied to the plants through drippers.

Fertilizers

Fertilizer dosage has to be dependent on growing media. Soilless mixes have lower nutrient holding capacity and therefore require more frequent fertilizer application. Essential elements are at their maximum availability in the pH range of 5.5 to 6.5. In general Micro elements are more readily available at lower pH ranges, while macro elements are more readily available at pH 6 and higher.

Forms of inorganic fertilizers

Dry fertilizers, slow release fertilizer and liquid fertilizer are commonly used in green houses.

Slow release fertilizer

They release the nutrient into the medium over a period of several months. These fertilizer granules are coated with porous plastic. When the granules become moistened the fertilizer inside is released slowly into the root medium. An important thing to be kept in mind regarding these fertilizers is that, they should never be added to the soil media before steaming or heating of media. Heating melts the plastic coating and releases all the fertilizer into the root medium at once. The high acidity would burn the root zone.

Liquid fertilizer

These are 100 per cent water soluble. These comes in powdered form. This can be either single nutrient or complete fertilizer. They have to be dissolved in warm water.

Fertilizer Application Methods

1. Constant feed

Low concentration at every irrigation are much better. This provides continuous supply of nutrient to plant growth and results in steady growth of the plant. Fertilization with each watering is referred as fertigation.

2. Intermittent application

Liquid fertilizer is applied in regular intervals of weekly, biweekly or even monthly. The problem with this is wide variability in the availability of fertilizer in the root zone. At the time of application, high concentration of fertilizer will be available in the root zone and the plant immediately starts absorbing it. By the time next application is made there will be low or non-existent. This fluctuation results in uneven plant growth rates, even stress and poor quality crop.

Fertilizer injectors

This device inject small amount of concentrated liquid fertilizer directly into the water lines so that greenhouse crops are fertilized with every watering.

Multiple injectors

Multiple injectors are necessary when incompatible fertilizers are to be used for fertigation. Incompatible fertilizers when mixed together as concentrates form solid precipitates. This would change nutrient content of the stock solution and also would clog the siphon tube and injector. Multiple injectors would avoid this problem. These injectors can be of computer controlled H.E. ANDERSON is one of the popular multiple injector.

Fertilizer Injectors

Fertilizer injectors are of two basic types: Those that inject concentrated fertilizer into water lines on the basis of the venturi principle and those that inject using positive displacement

A. Venturi Principle Injectors

1. Basically these injectors work by means of a pressure difference between the irrigation line and the fertilizer stock tank.

a) The most common example of this is the HOZON proportioner.

b) Low pressure, or a suction, is created at the faucet connection of the Hozon at the suction tube opening. This draws up the fertilizer from the stock tank and is blended in to the irrigation water flowing through the Hozon faucet connection.

c) The average ratio of Hozon proportioners is 1:16. However, Hozon proportioners are not very precise as the ratio can vary widely depending on the water pressure.

d) These injectors are inexpensive and are suitable for small areas. Large amounts of fertilizer application would require huge stock tanks due to its narrow ratio.

B. Positive displacement injectors.

1. These injectors are more expensive than Hozon types, but are very accurate in proportioning fertilizer into irrigation lines regardless of water pressure.
2. These injectors also have a much broader ratio with 1:100 and 1:200 ratio being the most common. Thus, stock tanks for large applications areas are of manageable size and these injectors have much larger flow rates.
3. Injection by these proportioners is controlled either by a water pump or an electrical pump.
4. Anderson injectors are very popular in the greenhouse industry with single and multiple head models.
    a. Ratios vary from 1:100 to 1:1000 by means of a dial on the pump head for feeding flexibility.
    b. Multihead installations permit feeding several fertilizers simultaneously without mixing. This is especially significant for fertilizers that are incompatible (forming precipitates, etc.) when mixed together in concentrated form.
5. Dosatron feature variable ratios (1:50 to 1:500) and a plain water bypass.
6. Plus injectors also feature variable ratios (1:50 to 1:1000) and operates on water pressure as low as 7 GPM.
7. Gewa injectors actually inject fertilizer into the irrigation lines by pressure.
    a. The fertilizer is contained in a rubber bag inside the metal tank. Water pressure forces the fertilizer out of the bag into the water supply.

b. Care must be taken when filling the bags as they can tear.

c. Ratios are variable from 1:15 to 1:300.

8. If your injector is installed directly in a water line, be sure to install a bypass around the injector so that irrigations of plain water can be accomplished.

General problems of fertigation

Nitrogen tends to accumulate at the peripheries of wetted soil volume. Hence, only roots at the periphery of the wetted zone alone will have enough access to Nitrogen. Nitrogen is lost by leaching and denitrification. Since downward movement results in permanent loss of NO3 –N, increased discharge rate results in lateral movement of N and reduces loss by leaching.

Phosphorous

It accumulates near emitter and P fixing capacity decides its efficiency. Low pH near the emitter results in high fixation.

Potassium

It moves both laterally and downward and does not accumulate near emitter. Its distribution is more uniform than nitrogen and phosphorous.

Micronutrients

Excepting boron, all micronutrients accumulate near the emitter if supplied by fertigation. Boron is lost by leaching in a sandy soil low in organic matter. But chelated micronutrients of Fe, Zn can move away from the emitter but not far away from the rooting zone.

Sprinkler systems

Sprinkler irrigation is a method of applying irrigation water which is similar to natural rain.In this way the whole surface is irrigated. The spray lines are placed high in the greenhouse, above the crop canopy. The plants are wetted with each irrigation.

Advantages

- The whole surface and root zone is wetted.
- Fertigation can be included, but proper planning during the irrigation is necessary to avoid nutrient precipitation on the leaves.

Disadvantages

- The plants become wet each time.
- The irrigation intensity is relatively high, short irrigation rounds are technically difficult and lead to high heterogeneities, hence leaching is a risk.
- Irrigation with sprinklers allows large cropped areas in glasshouse or polytunnels.
- Anti-drip nozzle which hold the water in the lines should be used to avoid any dripping onto the crop after switched off the system.
- Fogging or humidification nozzle can also be used instead of sprinkler to irrigate different crops. This is the ideal solution for cooling and humidifying of the greenhouse.
- A boom system consists of one or more pipes containing nozzles that apply water as the system moves over the plants.
- It may be suspended from an overhead rail system. Water is supplied by a battery pack or electric supply cable. Each boom system is independently designed to the growers need.

Strip irrigation by mini sprinklers and sprayers

Irrigation water is applied only as small strips or as only one plant row. The system is placed at the soil surface keeping the crop dry during the irrigation.

Sub irrigation systems

These systems deliver water and nutrients at the base of containers and dampening of soil/substrates arises from capillarity. These are closed systems that recirculate water from central reservoir to benches (e. g. ebb-and-flow, trough) and flooded floor systems, and the solution is drained back to the reservoir after irrigation. Collection trays and capillary mats re also efficient sub irrigation systems.

Irrigation scheduling

Irrigation scheduling methods can be based on

Soil monitoring

Plant monitoring

Soil-water balance

Simulation

Irrigation and fertigation: As compared to open field conditions, efficient water usage can be achieved through protected cultivation. Drip irrigation technology has come to stay in greenhouse production systems and it not only helps in using water efficiently but also can be responsible for reducing diseases that develop in rather moist conditions. Fertigation, i e irrigation combined with fertilizer application, requires consideration of plant needs, soil properties and technological requirements. It allows a precise and homogeneous application of nutrients in the area where the active roots are concentrated. The high potential efficiency of fertigation results from the possibility of using the optimal concentration of nutrients and a high density of roots in the moistened soil zone. The understanding of plant growth behavior including nutrient requirements and rooting patterns, soil chemistry (solubility and mobility of the nutrients), fertilizers chemistry (mixing compatibility, precipitation, clogging and corrosion) and water quality factors (pH, salt and sodium hazards) is required for effective fertigation. The quality of greenhouse vegetable crop improves in all respects such as fruit size, TSS content, ascorbic acid content and pH. With fertigation through the drip irrigation system, crop foliage can be kept dry thus avoiding leaf burn and delaying the development of plant pathogens. Depending upon different varieties, season and crop stage fertigation schedule for different crop varies and standardized at local level.

Soil Moisture Sensors

Since the direct gravimetric measurement of free soil moisture requires removing, drying, and weighting of a soil sample, soil moisture sensors measure the volumetric water content indirectly by using some other property of the soil, such as electrical resistance, dielectric constant, or interaction with neutrons, as a proxy for the moisture content.

Potential benefit of soil moisture sensors

- Provide feedback on soil moisture level for correct scheduling of irrigation.
- Provide additional information on soil environment e.g. soil temperature, Electrical conductivity (EC).
- Determine effectiveness of irrigation-monitor wastage through deep drainage
- Evaluate or analyse plant demand for water
- Detect leaching of water

- Switching of irrigation controller
- Save water by improving irrigation effectiveness
- Reduce water wastage

Sensing Techniques

Volumetric

Water content sensors (WC) measure the total water held in the soil i. e. water locked into the soil structure plus available water)

Tensiometric

Matric potential sensors measure the effort required to extract the water from the soil.

Volumetric Methods

- All volumetric water content sensors estimate the volume of water in a sample volume of undisturbed soil.
- The quantity is useful for determining how saturated the soil is (i. e. fraction total soil volume filled with the soil aqueous solution).
- When it is expressed in terms of depth (i. e. volume of water in soil down to a given depth over a unit surface area, mm of water), it can be compared with other hydrological variables like precipitation, evaporation, transpiration, deep drainage etc.

Neutron Moderation

- Fast neutrons are emitted from a radioactive source (241Am/9Be) and when they collide with particles having the same mass as a neutron (i.e. protons, H+), they slow down dramatically, building a "cloud "of "thermalized" (slow down) neutrons.
- Since water is the main source of hydrogen in most soils, the density of slowed down neurons formed around the probe is nearly proportional to the volume fraction of water present in the soil.

Dielectric Methods

Dielectric sensors(or Capacitance sensor) use capacitance to measure the dielectric permittivity of a surrounding medium. The configuration is like the neutron probe where an access tube made of PVC is installed in the soil; probes can also be modular (comb-like) and connected to a logger. The sensing head consists of an oscillator circuit, the frequency is determined by an annular electrode, fringe-effect capacitor, and the dielectric constant of the soil. Each capacitor sensor consists of two metal rings mounted on the circuit board at some distance from the top of the access tube. These rings are a pair of electrodes, which form the plates of the capacitor with the soil acting as the dielectric in between. The plates are connected to an oscillator, consisting of an inductor and a capacitor. The oscillating electrical field is generated between the two rings and extends into the soil medium through the wall of the access tube. The capacitor and the oscillator form a circuit, and changes in dielectric constant of surrounding media are detected by changes in the operating frequency. The capacitance sensors are designed to oscillate in excess of 100 MHz inside the access tube in free air. The output of the sensor is the frequency response of the soil's capacitance due to its soil moisture level.

29

Drainage

Drainage is the natural or artificial removal of excess water from a surface or sub-surface from an area or field or soil or greenhouse. The internal drainage of most agricultural soils in greenhouse is good enough to prevent severe waterlogging. Water logging creates anaerobic conditions that harm root growth in absence of oxygen. Soils need artificial drainage to improve production or to manage water supplies in certain conditions. Though the roots need moisture but they need oxygen too and poor drainage is g result in roots rotting and death of plants.

The irrigation water inside the greenhouse may accumulate and cause anaerobic condition. Similarly plants in pots should not accumulate water. Water should not stand in pots and beds for longer period. If it stays longer drainage is poor and needs to be corrected. It is better to have good drainage system in greenhouses or protected structures.

Besides irrigation rains water seeps in the greenhouse. The drainage system prevents running or its accumulation in low lying portion of the greenhouse. Drainage system in greenhouses is a must for avoiding damage to the crops.

Common types of drainage

- Ditches
- Subsurface drain lines (tile)
- Mole drains
- Surface drains
- Raised beds and ridges

Clay soils may also require surface drainage, which involves shaping the land to allow water to discharge over the soil surface to the edge of fields, where it can enter waterway. Soil shaping is also used to smooth out localized depressions where water would otherwise accumulate and remain ponded for extended periods of time.

A very modest system of drainage involves the use of ridges and raised beds, especially on fine-textured soils. This involves limited surface shaping, in which the crop rows are slightly raised relative to the inter-rows. This may provide a young seedling with enough aeration to survive through a period of excessive rainfall. These systems may also include reduced tillage—ridge tillage involves minimal soil disturbance—as well as controlled traffic to reduce compaction. Sze and shape of the beds and pots and the soil mixture inside should facilitate the drainage of excess soil water below the root zone area.

In greenhouses major problem is drainage of rain water. The runoff created due to rain fallen on a greenhouse is almost 100 percent of the quantity of the rain received. As the roof of the greenhouses is generally provided with sufficient slope and less friction, the amount of rain water accumulated for immediate disposal is very high necessitating immediate disposal or draining. Without proper drainage system for quick disposal of rain water from the greenhouses premises, the water may enter the greenhouse through seepage or otherwise and damage the greenhouse crop. Proper drainage system for greenhouse cultivation should consist of

1. Roof water disposal system attached with greenhouse structure equipped with gutter rainwater pipes.
2. Open or underground drains are essential to collect and dispose of the rain water in a storage or water harvesting tank or structure.
3. Main drainage channel leading towards safe disposal area.

In general, the drain pipes are connected to gutters and drains situated outside the greenhouse. The drains are completely seepage and leakage proof which are connected to water storage/harvesting structure. The slope and size of gutter drain pipes and surface drains should be designed in such a way that it can quickly dispose of the peak run off water.

Besides, the open area existing in between and surrounding the greenhouse should be developed in such a way that the rain water is collected either in water harvesting tank or drained of far away from the greenhouse by earthen channels.

30

Flooding and Leaching

Flooding

Flooding affects both above and below ground ecosystem processes in greenhouses. Below-ground changes may be less obvious, they are as important as the above-ground changes. Soil microorganisms are sensitive to disturbance, and shifts in soil microbial community structure are expected when anaerobic conditions develop from flooding. The useful aerobic microbes are destroyed by flooding. It takes time to built-up their population in the soil to benefit plants. Due to anaerobic condition created by flooding in greenhouse plant roots in soil are not able to breath, function and hence wilting of plants occurs. Flood irrigation in excess create flood like condition in greenhouses at time. Excess water in soils is a major factor affecting plant survival and functioning. In saturated soils, the supply of atmospheric oxygen into the soil is curtailed and various facultative and obligate anaerobic microorganisms use oxidized compounds as electron acceptors for respiration, thus converting them to reduced forms. The reduction and the associated processes influence plant survival, growth and functioning in wetlands. Flood irrigation should always be just wet the soil not flood inviting stagnation of water.

A chain of reactions is initiated upon soil flooding leading to reduced soil redox potential conditions. These reactions include physical, chemical and biological processes that have significant implications on plants growth. Physical processes include restriction of atmospheric gas diffusion in the soil leading to depletion of soil oxygen and accumulation of carbon dioxide. Shortly after flooding, the limited supply of oxygen in soil pore spaces is depleted rapidly by roots, microorganisms, and soil reductants. This process leads to oxygen depletion and reduction in soil oxidation reduction potential followed by a chain of soil chemical changes. The processes that follow include denitrification, reduction of iron, manganese and sulphate, and changing soil pH

Many terms such as “flooded”, “saturated”, “waterlogged describe oxygen-deficient root medium-soil. Obviously, these terms are qualitative and do not quantitatively define the rhizosphere (root zone). Furthermore, methods used to quantify oxygen content and oxygen diffusion rate in drained soils cannot be employed effectively in wet soils. Plant nutrition is influenced by soil flooding and the processes associated with reducing soil redox potential condition. Many factors including soil physicochemical characteristics, nutrient pools, plant developmental and physiological status, and flood-tolerance capabilities are important. Several nutritional deficiencies and toxicities may occur. Reduced soil conditions may lead to inhibition of nutrient uptake and transport due to root dysfunction, death and blockages in the vascular systems resulting from phyto-toxin damages. Oxygen supplies to the roots are critical for nutrient uptake and ion transport. Flood-induced stress includes reduced water uptake due to root dysfunction that results from soil O2-deficiency including altered cation and anion concentration in xylem sap as well as decreased hydraulic conductance of roots. Reduction of water and nutrient uptake, disturbance of hormonal balance such as a decrease in gibberellin and cytokinin and an increase in abscisic acid and ethylene in response to flooding.

The severity of oxygen loss and the effects of reduction intensity and capacity on plant functioning due to flooding are common but may vary across plant species or crop varieties and hybrids. By flooding the greenhouse soil without crop can be useful in disinfestation if soil pathogens have multiplied to a dangerous level.

Leaching

Leaching is referred tothe loss of water-soluble plant nutrients from the soil due to rain and irrigation or applying a small amount of excess irrigation to avoid soil salinity. Soil structure, crop planting, type and application rates of fertilizers, and other factors are taken into account to avoid excessive nutrient loss. Leaching may also refer to the practice of applying a small amount of excess irrigation where the water has a high salt content to avoid salts from building up in the soil (salinity control). Where this ispracticed, drainage must also usually be employed, to carry away the excess water.

There are two types of leaching in the garden: Leaching of soil The soil in your garden is like a sponge. When rain falls, the soil near the top absorbs as much as possible, keeping the moisture available to the plants growing there. Once the soil is filled with all the water it can hold, the water begins to leak downward through the layers of rock and subsoil beneath your garden. When the water sinks down, it takes soluble chemicals with it, such as nitrogen and other fertilizer components, as well as any pesticides you may have used. This is the first of the types of leaching.

What soil type is most prone to leaching? The more porous the soil, the easier it is for chemicals to pass through. Pure sand is probably the best leaching type, but isn't very hospitable to garden plants. In general, the more sand your garden soil has, the more likely it is that you will have excess leaching. On the other hand, soil with more of a clay component presents less of a leaching problem. Leaching in plants is more an environmental concern than that of poor drainage. Once your pesticides have leached from the plants themselves down through your soil into the water table, they begin to affect the environment. This is one reason why many gardeners prefer organic methods of pest control.

Leaching of potted plants: Leaching in plants can happen in potting containers. Once the chemicals have drained down through the soil, they can leave a crust of soluble salts on the surface, which makes it hard for the soil to absorb water. Removing this crust with water is the other type of leaching. Leaching garden plants grown in containers is the process of washing the salts from the surface of the soil. Pour large amounts of water through the soil until it runs freely from the bottom of the planter. Leave the container alone for about an hour, then do it again. Repeat the process until you don't see any more white covering on the surface of the soil.

Leaching of nitrogen

Nitrogen is a common element in nature and an essential plant nutrient. Approximately 78% of Earth's atmosphere is nitrogen (N_2). The strong bond between the atoms of N_2 makes this gas quite inert and not directly usable by plants and animals. As nitrogen naturally cycles through the air, water and soil it undergoes various chemical and biological transformations. Nitrogen promotes plant growth. Livestock then eat the crops producing manure, which is returned to the soil, adding organic and mineral forms of nitrogen. The cycle is complete when the next crop uses the amended soil. To increase food production, fertilizers, such as nitrate (NO_3) and ammonium (NH_4), which are easily absorbed by plants, are introduced to the plant root zone. However, soils do not absorb the excess NO_3 ions, which then move downward freely with drainage water, and are leached into groundwater, streams and oceans. The degree of leaching is affected by:

- soil type and structure for example, sandy soil holds little water while clay soils have high water-retention rates;
- the amount of water used by the plants/crops;
- how much nitrate is already present in the soil?

The level of nitrous oxide (N_2O) in the Earth's atmosphere is increasing at a rate of 0.2 to 0.3% annually. Anthropogenic sources of nitrogen are 50% greater than from natural sources, such as, soils and oceans. Leached agricultural inputs, i.e. fertilizers and manures, accounts for 75% of the anthropogenic source of nitrogen.

Phosphorus (P) Leaching

Phosphorus (P) is a key nutrient regarding the eutrophication of surface waters and has been shown to limit algae growth in lake environments. Loss of P from agricultural fields has long been recognized as one of the major threats to surface water quality. Leaching is an important transport pathway for P losses from agricultural fields in predominantly flat areas with sandy soils or soils prone to preferential flow. As opposed to nitrogen phosphorus does interact with soil particles via adsorption and desorption. Important potential adsorption sites for P in soils are surfaces of iron and aluminium oxides or hydroxides such as gibbsite or ferrihydrite. Soils, especially those rich in such minerals do hence have a potential to store P added with fertilizer or manure. Adsorbed P stands in a complex equilibrium with P in the soil solution which is controlled by a multitude of different factors such as:

- the availability of adsorption sites
- the concentration of phosphorus and other anions in the soil water solution
- soil pH
- soil redox potential

Phosphorus will leach when this equilibrium is shifted such that either previously adsorbed P is released into the soil solution or additional P cannot be adsorbed anymore. Many cultivated soils have been receiving fertilizer or manure P in amounts frequently exceeding the crop demand and this often over decades. Phosphorus added to such soils leaches simply because most of the potential adsorption sites are occupied by P input from the past, so called legacy phosphorus. Leaching of P may also be caused by changing chemical conditions in the soil. A decrease in the soils redox potential due to prolonged water saturation may lead to reductive dissolution of ferric iron minerals that are important P sorption sites. Phosphorus adsorbed to these minerals is consequently released into the soil solution as well and may be leached. This process is of special concern in the restoration of natural wetlands that have previously been drained for agricultural production.

Calcium and magnesium leaching is also observed in greenhouses. Leaching of N, Ca and Mg occurs more in loam soils. Leaching of nutrient can be minimized by adding appropriate organic matter to the soil. Soils are to be amended with suitable organic matters to arrest leaching of useful plant nutrients from soil. Micro-irrigation in greenhouse crops would reduce leaching of useful macro and micronutrients. Nutrients require water as a medium to go down the soil and pollute the groundwater.

31

Soil Pasteurization in Peat Moss and Mixtures

Pasteurization: The heat treatment given to soil to free it of harmful microorganisms or pathogens and pests is called pasteurization. It kills weed seeds also.

Pasteurization of greenhouse plant growing media

Greenhouse growing medium may contain harmful pests such as disease causing organisms, nematodes, insects and weed seeds. It is advisable to decontaminate it by heat treatment or by treating with volatile permissible chemicals.

Agent	Method	Recommendation
Heat	Steam	30 min at 82.22 °C
Methyl bromide	10 ml/cu. ft. of medium	Cover with gas proof cover for 24-48 hr. Aerate for 24-28 hr before use.
Chloropicrin	(Tear gas) 3-5 ml/cu. ft. of medium	Cover for 1-3 days with gas proof cover after sprinkling with water. Aerate for 14 days or until no odour is detected before using.
Basamid	8.0 g/cu.ft. of medium	Cover for 7 days with gas proof cover and aerate for atleast a week before use.
Formalin	20 ml/l of water (37%)	Apply 2 l/cu.ft. cover for 14 to 36 hr and aerate for at least 14 days.

Disinfection of the growing media can also be achieved by fungicides or bactericides Use only permissible disinfectants.

Fungicides and their effect on a few fungi

Chemical	Rate of application	Effect against
Captan	2 g/l of water	*Pythium, Fusarium, Rhizoctonia and Phytophthora. Some extent to root and stem rot, white mold,black rot, crown rot and damping off.*
Metalaxyl + Mancozeb (Ridomil MZ 72 WP)	1 g/l of water	*Pythium, Phytophthora, Fusarium* and other soil borne pathogens

Temperature necessary to kill soil pests

- 46°C for water moulds (*Pythium* and *Phytophthora*)
- 49°C for nematodes
- 57°C for worms, slugs and centipedes
- 60°C for most plant pathogenic bacteria
- 71°C for soil insects
- 82°C for most of weed seeds
- 93°C for few resistant weed seeds and plant viruses

32

Rock Wool and Other Inert Media

Soilless media or culture

Gravel culture is a general term which applies to the growing of plants without soil in an inert medium into which nutrient solutions are usually pumped automatically at regular intervals. Haydite (shale and clay fused at high temperatures), soft- or hard-coal cinders, limestone chips, calcareous gravel, silica gravel, crushed granite and other inert and slowly decomposing materials are included in the term "gravel". The more important greenhouse flowering crops include roses, carnations, chrysanthemums, gardenias, snapdragons, lilies, asters, pansies, annual chrysanthemums, dahlias, bachelor buttons and others.

Media ingredients and Mix

Commercially available materials like peat, sphagnum moss, vermiculite, perlite and locally available materials like sand, red soil, common manure/ compost and rice husk can be used in different proportions to grow greenhouse crops. These ingredients should be of high quality to prepare a good mix. They should be free from undesirable toxic elements like nickel, chromium, cadmium, lead etc. Most common medium other than soil used in India in greenhouses is coco peat. In other countries these are Rockwool and Peat-vermiculite.

Now coconut coir (CC) has been increasingly used as an alternative to RC and peat because of environment concerns. Besides being organic and environment friendly material it has stable physicochemical and biological properties. CC is the coconut waste consisting of the dust and short fibres and approximately 12 million tonnes are produced annually in the world. Due to its good water retention and aeration characteristics, CC has gradually become the most potential alternative to both RC and peat in substrate cultivation.

In substrate cultivation, crops were planted in a small volume of growing media, resulting in limited nutrients and water for root absorption. Hence, mineral nutrient management is a key factor determining the yield and nutritional quality of vegetable and other crops during substrate cultivation. Generally, the retention, movement and availability of mineral nutrients in root-zone are related to several properties of a substrate, such as particle size, water and nutrient holding capacities, cation exchange capacity and nutrient content.

Table: Selected physical and chemical properties of Rockwool, coir and peat-vermiculite.

Properties	Rockwool	Coir	Peat-vermiculite
EC (dS m^{-1})	0.06	0.10	1.10
pH	6.5	6.1	7.1
C (%)	2.2	49.5	15.9
N (mg kg^{-1})	56	44	64
P (mg kg^{-1})	30	38	42
K (mg kg^{-1})	178	1560	246
Ca (mg kg^{-1})279581668	279	58	1668
Mg (mg kg^{-1})	216	55	636
S (mg kg^{-1})	303	405	645
Water porosity (%)	84.180.0	80.0	64.4
Air porosity (%)	5.1	5.6	2.0
Bulk density (g cm^{-3})	0.08	0.2	0.4

(Front Plant Sci. 2017; 8: 1327)

Desirable nutrient level in greenhouse growth media

S.No.	Category	Concentration (mg/l)			
		NO3	N	P	K
1.	Transplants	75	125	10-15	250-300
2.	Young pot & foliage plants	50	90	6-10	150-200
3.	Plants in beds	125	225	10-15	200-300

There are several other inert materials like moss, both organic and inorganic for soilless production of crops. They are locally available at a cheaper rate and can be used either alone or as a mix.

33

Nutrient Film Technique (NFT)

Nutrient film technique popularly known as NFT is one of the methods of soilless farming or hydroponics where in plants have their roots in a shallow stream of recirculating water containing all the nutrients required for plant growth dissolved without any solid growing medium. The lower portion of roots remain submerged in flowing nutrient medium in a water tight gully having enough air for the roots. The nutrient film technique was developed during the late 1960s by Dr Allan Cooper at the Glasshouse Crop Research Institute, England. The method is already used in commercial horticulture in many countries including India. NFT is potential for other crops such as animal fodder and even energy-source plants for fuel production.

NFT related technologies

Hydroponics : Cultivation of plants in nutrient solutions is known as hydroponic cutlivation. This ia gerally done under protected structures and in containers arranged horizontally or vrtically (vertical garden). Use of nutrient film or fertilizers is made to feed the plants. Hydroponics can be practiced following nutrient film technique method, trough method, deep water floating method and coco peat bag culture.

Advantages of hydroponics

One can grow plants anywhere. It uses 20 times less water than soil based cultivation and 20 per cent less space for growing. It requires less of water and pesticides. Water and waste nutrients can be reused. No intercultural operation like weeding, tilling, mulching, change of soil etc. are required

Disadvantages of hydroponics

Putting together a hydroponic system is expensive. Constant supervision is required. Hydroponic systems are vulnerable to power outages. Growing a hydroponic garden demands technical expertise. Large scale production yet to be standardized.

Aeroponics: This is a form of hydroponics. Nutrient and water mixture is sprayed in caged roots under dark at certain intervals with the help of sensors. This conserves water, space and is less cumbesome. It demands constant supervision and has disadvantages as listed under hydroponics.

Aquaponics : Involves production of fish and vegetables with same input resource of water and nutrients. Feed is given to the fishes in the pond not to the vegetables or plants grown as hydroponics. Fish pond or tank water is used as hydroponics or areoponics with or without additional fertigation. This results in effective utilization of fish pond water and its nutrients for production of vegatables/plants simultaneously with added returns on input resources. Plants act as biofilter for the recycled water to fish pond.

The system

A properly designed NFT system is based on using the right channel slope, the right flow rate, and the right channel length. The plant roots are exposed to adequate supplies of water, oxygen and nutrients- all essential for plant roots growth and their functions.

NFT, because of its design, provides a system wherein all three requirements for healthy plant growth can be met at the same time, provided that the simple concept of NFT is always remembered and practiced. The result of these advantages is that higher yields of high-quality produce are obtained over an extended period of cropping. A drawback of NFT is that it has very little buffering against interruptions in the flow of nutrients and water, such as a result of a power failure. Big NFT farms have their back power supply to overcome this problem. Another problem can be of continuous flow of the nutrient which is facilitated by providing 1:30 or 1:40 slopes in channels. It would be advisable to provide slope on the floor. A properly designed NFT system is based on using the right channel slope, the right flow rate, and the right channel length Chances of disease spread through nutrient solutions should be taken care of. It is, in general, one of the more productive techniques.

In NFT the nutrient is pumped into the growing tray/trough (usually a tube) and flows over the roots of the plants and flows back into the reservoir. The reservoir is to be checked for nutrient solution, TDS (total dissolved solids), pH and EC (electrical conductivity). There is no growing medium used other than air necessitating replacing the growing medium after every crop. It is a big advantage. Normally the plant is supported in a small plastic bucket with the roots dangling into the nutrient solution.

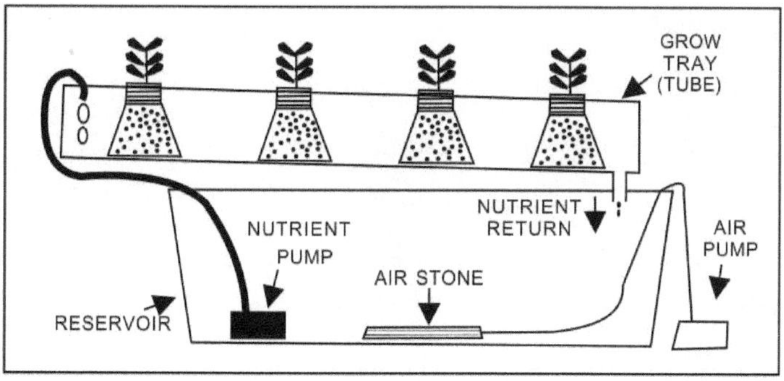

NFT, because of its design, provides a system wherein all three requirements for healthy plant growth can be met at the same time, provided that the simple concept of NFT is always remembered and practiced. The result of these advantages is that higher yields of high-quality produce are obtained over an extended period of cropping. NFT remains a very popular system for home use. NFT ensures that plants have unlimited access to water at all times, but it is now observed that fruiting vegetable crops (tomatoes, capsicums and cucumbers) can benefit from carefully limited required water supplies not the continuous water supply. Leafy vegetable crops like lettuce benefit from unlimited water supplies and are widely grown using NFT world over.

NFT system is comprising of three components

A. Layout and equipment

B. Operation of the NFT system involving formulation of nutrition solutions, measuring pH, EC and TDS, water, climate and crop protection and working temperature

C. Use of NFT for various Crops-Leafy vegetables and certain herbs do well.Fruit vegetables can be grown but they do well with inert medium like Rockwool, coco peat etc

After each crop your NFT channels should be cleaned to prevent any pathogens moving on to your next crop.

Aquaponics where fish are fed with their feed and their excreta is circulated in leafy vegetable growing NFT system is economical as there is saving on nutrients and water for vegetables in aquaculture and vegetable NFT separately. Whether it is used exclusively or in conjunction with another hydroponic system, NFT will continue to have a place in the future of hydroponics and indoor horticulture. The indoor horticulture in certain crops NFT has proved better than other hydroponic systems.

Advantages of NFT

1. Easy to examine the health of roots
2. Saving in nutrient and water consumption by plants
3. Chances of contamination of ground water are minimized, hence environment friendly.
4. Ease in built –up and maintenance of system
5. Comparatively inexpensive system
6. Growing medium not required.
7. Due to constant flow no need of aeration of reservoir and roots.
8. No fuss of timer or watering cycle.

Disadvantages

1. Stoppage of flow would lead to dry out of roots.
2. Vigorously growing roots can block the flow of nutrients.
3. Pump failure may give setback to plants.
4. Large tap root plants cannot be grown in this system.

34

Problems / Constraints of Greenhouse Cultivation and Future Strategies

The popularity of greenhouse-grown produce continues to rise as it provides a variety of produce during the "off" season. Greenhouses are good for healthy plant nursery production and other a number of crops for profit. Though the greenhouse farming in India like in other countries is progressing but there are certain problems and constraints as stated below.

1. Greenhouse production generally is highly technical and requires constant and continuous attention during cropping period. Relevant information about the crop environmental requirement such as right temperature, humidity etc is essential.

2. Knowledge about the right type of inputs for greenhouse production is a must which is not easily available.

3. Greenhouses cultivation increases the land, water and fertilizer productivity along with reduction in pesticide use. But such interventions are associated with very high initial cost. Although, it is linked with several Government Schemes in India and have 50-90% of subsidies depending on intervention, status / socio-economic condition and policy of State and Central Government.

4. Uninterrupted and regular power supply is required for operating cooling and heating system of the controlled environment greenhouses which necessitate a back-up power pack which is also expensive and underutilized.

5. Cladding material of required quality and specification is not readily or easily available.

6. Non-availability of tools and implements for facilitating crop-production operations under greenhouse.

7. There is a lack of major research program on greenhouse crop production in the country. Region specific technologies are needed in absence of controlled environment greenhouses.

8. Inadequate or scanty breeding work programs for development of suitable varieties/ hybrids of crops for greenhouse cultivation. Exotic seeds are very costly and are out of reach of the Indian growers. Their regular supply is uncertain.
9. There is a problem of pollination in cross pollinated greenhouse crops
10. There is an increasing threat of bio-stresses (insects and diseases) in greenhouse cultivation particularly root-knot nematodes and *Fusarium* wilt. Use of pesticides is not desirable keeping in view the market and environment.
11. Strong winds and storms are a big threat to greenhouse industry.
12. In India greenhouse research is quite inadequate for claddings of greenhouses, automation and mechanization gadgets even the seeds of suitable crops are imported which makes it unsustainable.
13. Loss of soil fertility is a common issue as majority of greenhouses after three years of cultivation. Disease and pests multiply and requires frequent spraying of pesticides which is not only expensive but nor environment and consumer friendly.
14. Crop rotation becomes difficult due to continuous demand of a particular crop in the market.
15. Methods of growing crops in soil-less cultures and pots are highly technical and requires better skill.
16. Soil disinfection by solarisation/ sterilization/fumigation of soil bed are essential but cumbersome process.
17. Greenhouse cultivation requires high hygiene of premises maintenance of clean farm tools and planting materials and water.

Management problems in greenhouse cultivation

The troubles which arise in the cultivation of crops in the greenhouse may be divided into several groups

a) failure to supply the essential factors for optimum growth such as light, moisture, carbon dioxide and heat in amounts necessary for each individual crop

b) fertilizer deficiencies

c) fertilizer excesses

e) attacks by insects-pests

Fertilizer deficiencies

Symptoms of deficiencies of various fertilizers in greenhouses cultivation are:.

Chlorosis

This is a term used to denote the loss of normal green colour from the foliage whether it is on the older, more mature leaves or the younger foliage. The entire leaf may be affected, or just areas between the veins, in which case the yellowing is most usually in irregular patches shading into the green colour. Sometimes only the margin of the leaf or leaflets may be yellow, while the centre of the foliage is almost a normal green.

Necrosis

This refers to the death of the area severely affected by chlorosis. Necrotic spots or areas can also be caused by spray or aerosol damage, sunscald and other such factors which may have no relation of fertilizer.

Nitrogen deficiency

Generally the entire plant becomes lighter green, but the effect will be most noticeable on the older foliage. Gradually the oldest leaves lose their green colour, and most plants become yellow. The flowers are smaller and may lack well-developed colour.

Phosphorus deficiency

A purplish coloration developing first on the underside of the petiole, or leaf stem, which spreads to the main veins of the leaf is characteristic of this deficiency.

Potassium deficiency

The margins of the leaves of the older foliage become yellow, and the chlorosis progresses toward the mid-portion of the foliage as this deficiency increases in severity. The older leaves may drop in extreme cases of deficiency. Certain fumigants may cause marginal burning or chlorosis, and sometimes droplets of spray or fumigant may result in spots or blotches of chlorotic or necrotic nature.

Calcium deficiency

In sand culture, a typical symptom is the development of short clubby roots followed in a matter of several weeks by their death. In many cases insufficient calcium is associated with a low pH of the soil.

Iron deficiency

This is a rather common trouble although an actual lack of iron may not be the primary cause. As iron deficiency becomes more intense, necrotic areas appear on scattered portions of the yellow coloured leaves and the affected foliage may drop. Iron can become deficient in soil, but often the symptoms of this deficiency are induced by other causes from injury to the roots by over-watering or over fertilization. Nematodes, or other soil pests interfering with root growth can also induce iron chlorosis symptoms.

Boron deficiency

The number of cases where this is a limiting factor are few, and most of them are with certain rose and carnation varieties. The new foliage is thick or leathery and quickly becomes chlorotic. The rose flowers are usually very malformed. The stem tip dies, giving rise to growth of shoots immediately below, which in turn die at the tip, and a 'witches broom' effect is observed. Because deficiency symptoms can sometimes be confused with the effects of some other environmental factor of cultural practice, a thorough review of fertilizer application, soil testing, soil type, watering practices, and other procedures is warranted before hasty conclusions are reached.

Fertilizer excesses

An unfortunate belief among many growers is that when a plant does not grow under apparently favourable conditions, the trouble can be overcome by applications of fertilizer. This practice has resulted in untold damage or loss of crops, as more often than not the original trouble could have been too much fertilizer in the soil. If additional fertilizer is applied when no more is needed, the results can be very injurious. Sometimes the difference between a high but safe nutrient level and an injurious nutrient level is not very great and the margin of safety may be extremely small. Therefore, to test the soil in case of doubt to determine the advisability of fertilizer application.

Nitrogen excess

The plants exhibit heavy, rank growth, with large, dark green leaves that are often crisp and break easily. Additional nitrogen may inhibit root action, causing typical symptoms of iron chlorosis. If the root system is killed, the plants wilt excessively and never recover. This yellowing of the top foliage is very common in chrysanthemums and snapdragons. Over 75 ppm of nitrates is not safe.

Phosphorus excess

Over doses of phosphorus precipitate the iron from the soil solution and make it insoluble and unavailable, causing iron chlorosis to develop. Over 25 ppm may cause trouble.

Potassium excess

Up to a certain point, excessive potash apparently is not injurious. Greater amounts inhibit root action and may cause chlorosis, wilting, or immediate death of the plant. Over 60ppm is dangerous.

Calcium excess

Usually the pH of the soil will rise when there is excessive calcium. This causes iron chlorosis in many plants and has been called over liming injury. Over 300 ppm is high.

Iron excess

In the normal pH range of soil, there is little danger of excessive iron since phosphorus or calcium will precipitate it from the soil solution. At pH 5.0 or lower, iron becomes very soluble, and on hydrangeas brown dots appear on the leaves, indicating iron is being precipitated as water vapour is lost by transpiration.

Sulphate excess

A low pH may often be characteristic of soils high in sulfates. Sometimes high sulfates are encountered at pH 6.5. Over 600 ppm is toxic to most plants.

Boron excess

This trace elements is often found in soil to which unleached cinders have been added or where boric acid has been applied indiscriminately. Certain water supplies may have rather high amounts of boron. On roses, the serrations on the margin of the leaflets of the lower leaves turn black and the remaining leaves turn yellow and drop.

Aluminium excess

This is not troublesome except on hydrangeas that are being blued. Roots are burnt, and the plants wilt.

Soluble salts excess

Too much fertilizers in the soil injures or kills roots, and plant growth is severely reduced.

Remedial measures for excess fertilizer

When fertilizer levels rise to the point where they become toxic, immediate steps must be taken to remove the excessive materials. Excessive nitrogen leaching with heavy watering and application of straw mulch Excessive phosphorus cannot be leached. Only by addition of lime or iron sulphate Excessive potassium leaching may wash some quantity, but in clay soils removal may be almost impossible. Excessive calcium Acidifying the soil excessive iron Raising the pH or by addition of phosphorus excessive sulphates Leaching and avoiding the use of sulphate forms of fertilizers excessive boron Water glass, or sodium silicate can be dissolved in water at the rate of 100 cubic centimetres per gallon and applied to the soil. Excessive aluminium raising the pH or by adding phosphorus excessive soluble salts leaching.

Insect- pests in greenhouse

Greenhouse crops grown all over the world are vulnerable to various diseases and pest attacks as the protected crops provide stable and favourable microclimates for development of pest populations, which often limit the success of this crop production system. The losses caused due to pests in greenhouses crops like tomato, okra, capsicum, gerbera, carnation, cucumber, lettuce, beans, etc. are tremendous. Crop losses are mainly due to arthropod pests like mites, whiteflies, thrips, leaf-miners, aphids and diseases caused by virus, fungi, bacteria, nematodes etc. Among these various, insects are of much importance and need to be managed properly so as to prevent the crop losses and increase yields. The increasing need for higher crop yields both in the field and in greenhouses brings with it problems linked to large monocultures and pest attacks. Chemical pest control has to be reduced owing to its unwanted effects on non-targeted organisms and pest resistance Thus, alternative and sustainable longlasting pest control methods are urgently needed to enhance the activity of beneficial organisms.

Prevention and management of bio-stress (insect-pests) in greenhouse cultivation like any other farming is of utmost importance. The major biotic stresses in indoor farming of crops as stated above include spider mites, thrips, aphids, whiteflies, fungal gnats (*Bradysia* spp.), powdery mildew, downy mildew, grey mould, root rot are common insect-pests and diseases especially where humidity are high and excessive fertilizers are provided. The management of these stresses is advocated under following three subsystems.

1. ***Prevention or avoidance***: Exclusion of the infested area or pots, troughs or planting trays etc. general hygiene of plants, workers and accessories and water used for irrigation is of importance. Good Agricultural Practices

(GAP) protocols are nowadays becoming popular and include strict preventive protocols as per buyer requirements.

2. ***Monitoring and regulation:*** Regular monitoring and surveillance of the pestilence during the crop is linked with suitable and electronic or e-pest surveillance followed up with strict regulatory and quarantine practices for the import and export related items- particularly, seeds and planting material. Tools like yellow sticky traps, delta traps and other modern devices are to be used. Monitoring of disease-spreading plant material has to be done strictly at all levels.

3. ***Intervention or control measures***

   (a) Mechanical controls like manual or physical destruction, removal, termination, pinching, pruning, burning and destruction of pest sites or propagules to avoid building of high pest populations.

   (b) Physical measures of control include management of temperature from seedling stage to harvesting, shading and cooling management, aeration, carbon dioxide enrichment, oxygenation of substrate for suppressing anaerobic pathogens, particularly fungi like Pythium, CO_2 enrichment for more effective photosynthesis.

   (c) Bio-rationales and biological, crop and pest specific can be used as per need.

   (d) Chemical control, these options are available but should sparingly be used with great care and caution.

   Integrated pest management of biotic stresses in greenhouse cultivation/ farming is desirable.

Opportunities in greenhouse cultivation systems

Keeping in view the increasing population, changing climate, decreasing land holdings, increasing pressure on natural resources i.e. land and water and higher productivity of high quality produce Indian agriculture has to adopt protected cultivation particularly in horticulture crops involving hi-tech horticulture. Protective cultivation ensures intensive cultivation yielding manifold. Promotion of protected cultivation will certainly help in creation of huge self-employments for unemployed educated youths in rural areas and add to their pride which is lacking now. Production of vegetable and cut flower crops under protected conditions not only provides high water and nutrient use efficiency but it can easily increase the productivity by 3-5 folds over open field cultivation of these crops under varied agro climatic conditions of the country. This technology has very good potential especially in urban and peri-urban areas adjoining to the major citieswhich is a fast growing market for fresh produce of the country. Protected cultivation is one of answer to most of the burning issues facing Indian Agriculture including adversities due to uncertain and varying climatic conditions, climate change, improper uses and low productivity of natural resources, nutritional security in topographic and climatic disadvantaged areas, environmental pollution due to pesticide use, etc.

Greenhouse/protected cultivation is potential technology to contribute to high productivity and reduction in inputs under different climatic conditions by adopting appropriate protective structure and improved crop production technologies mainly for horticultural crops. This technology would be essential for production of quality planting material of horticultural crops such plug seedlings, grafted vegetable seedlings, robot horticulture nursery production, hybrid seed multiplication. Off season production of vegetable crops is of importance under protected/greenhouse cultivation.

Research conducted at various locations by the SAUs, PFDCs, ICAR, DRDO and other organizations have revealed that, there is a vast scope of utilizing the protected/greenhouse cultivation technology for growing many vegetables, flowers and few fruit crops in different parts of the country. The studies conducted by Defence Research and Development Organization (DRDO) have proved that even in high altitude areas of Leh and arctic deserts, the greenhouse cultivation can assure fresh supply of vegetables and other crops. A number of opportunities exist for nursery men, commercial crop growers, seed producers and researchers in utilizing greenhouse for their benefit. In temperate and subtropical zones protected farming can easily be used for raising vegetable nursery in late winters which could be transplanted in early spring. This can advance the cropping by one to one and half month and thus, may provide remunerative price to the farmers. In plains green house may be utilized for year round propagation of many tropical and subtropical fruits which could prove a boon for nursery men. There is a demand of quality vegetable nursery in the vegetable growing areas. Year round cultivation of selected vegetable crops such as tomato, capsicum and cucumber is possible under protected environment with single/ multi crop in a year, which fetches off season higher price with quality produce and also with lower cost of cultivation and longer duration of crop.There is high demand of colored capsicum, parthenocarpic cucumber, cherry tomato and hydroponically grown leafy vegetables in the hotel industry and export market throughout year at very attractive market price of the produce. India has a vast scope for exporting cut -flowers such as Gerbera, Carnation, Lilium, Roses, herrbs and vegetables grown under protected/ greenhouse environment.

Future strategies for greenhouse cultivation

With the advancement in agricultural technologies and other technologies the future agriculture would be technology dominated. Technology like greenhouse technology would be largely adopted and it would be not only common technology but dependable due to climate change.

- Standardizing proper design of construction of green/poly houses preferably region wise including cost effective cladding and glazing material.
- Developing cost effective agro-techniques for growing different crops and lowering energy costs if any of the greenhouse environment management.
- Developing professional and skilled green/poly house manufacturers.
- Creation of awareness among farmers pertaining to the potential of greenhouse/protected production. Further strengthening in needed.
- Human resource development is scanty in the country both for structure development and maintenance as well as crop production technology on green houses. Strengthening of both faculty and infrastructure is needed.
- A mission on greenhouse cultivation in the country under Ministry of Agriculture, and Farmers' Welfare is needed encompassing hi-tech greenhouse technologies.
- Institutes on greenhouse/protected cultivation under different major climatic zones of country are needed to overcome some of the major constraints mentioned above.
- There is need of computerized control systems which maximize returns: These include time base/volume base/sensor based irrigation system, opening and closing of ventilators and side wall roll up curtains, retractable roof, CO_2 generator, climate control, temperature, humidity, heat radiation, control of EC, pH, ppm level of elements in irrigation water etc. as required to the plant. There is urgent need for touch screen and press button technologies.
- Development of cost effective agro-techniques for growing of different crops in the different types of greenhouses and lowering energy costs of the greenhouse environment management.
- Import of planting materials, structural designs and production technologies which are not relevant under Indian conditions should be discouraged and in turn emphasis should be given to develop indigenous crop F_1 hybrids/ varieties suitable for greenhouses so that seeds are made available to the

growers in time and at comparatively cheaper rates. Imported technology and stores are to be indigenized and made Indian farmers friendly before passing on to the growers.

35

Future Thrusts on Greenhouse in India

Greenhouse technology development has made good progress in India during the last two decades, and steps necessary for promoting greenhouse cultivation of flower and vegetable crops have been started. Energy-efficient greenhouse cultivation continues to be an area of active research and development, and this is sought to be achieved through precision equipment and protocols. While the gap between the demand and supply of most horticultural crops remains wide and the country plans to double the production of horticultural crops by 2012, the protected cultivation technology holds the key to meet the targets. It clearly emerges that greenhouse technology has multipurpose application for sustainable development of a hilly zone. During harsh winter which otherwise threatens the survival of human beings in these remote hilly areas, popularization and further improvements in greenhouse management could provide a sign of relief in solitude of white snow cover.

In India future thrust would be on following:

A. Vertical farming

Vertical farming generally refers to the growing of crops mostly vegetables, ornamentals, and herbs on stacks of indoor shelves using artificial light and nutrient solutions, without much sunshine and soil. Such farms are not dependent on seasons/controlled environment and have ability to enhance production round the year with little risk of crop failure. Further they give fresh quality produce without depending on favourable climate, healthy soil, high water consumption and above all saves on labour, a scarce commodity today. Vertical farming has potential to sustain ever increasing world population especially in the urban areas with nutritional supplement thus providing food security. Vertical production of mushrooms, hydroponic green fodder, some vegetables and fruits and even poultry birds are either already in vogue or at advanced stage. Vertical gardens in ornamental horticulture, a component of vertical farming are also known as green walls, living walls, bio walls or vertical garden. It is a free-standing space or part of a building that is partially or completely covered with attractive vegetation luxuriantly growing in an organic or inorganic medium and in some cases soil also.

Hydroponics and vertical farming address the need of safe and healthy, pesticide/ acaricides/ insecticide free, natural antioxidant rich produce with low carbon and water foot print.

Important Features of Vertical Farming

Vertical farms enable the producer to:

- Grow food 24 hours a day, 365 days a year
- Protect crops from unpredictable and harmful weather
- Re-use of water collected from the indoor environment
- Provide jobs for residents/communities
- Minimize use of pesticides/fertilizers/ herbicides
- Drastic reduction and dependence on fossil fuels
- Prevent crop loss from storage/shipping/ long transportation
- Stop agricultural runoff, water saving up to 90%

1. Innovations in Vertical Farming

Some of the following vertical farming innovations have potential to support, supplement and improve existing agriculture techniques and technologies.

1. Hydroponics – Growing plants without soil or soilless production.
2. Aeroponics – Growing plants with no soil and very little water and nutrients in the form of mist or fog.
3. Aquaponics – An ecosystem-one drop two crops- crop and fish farming together with more or less same inputs (feed to the fish).
4. Lokal – Serving fresh food right where it is being grown
5. AeroFarms– Another smart vertical farming innovation - the technology of growing greens without sun or soil under controlled environment.
6. Plantscapers – A building that provides food for its occupants.
7. VertiCrop – A sustainable farming technique for urban areas.
8. Modular Farms – Produce fresh produce virtually anywhere in the world.
9. Cubic Farming Systems – The next-generation sustainable farming system-working with conveyor rotation and automated nutrient delivery system.
10. Zip Grow – Vertical farming for the modern farmers involving conveyor rotation, automated nutrient delivery and LED lighting.

11. Bowery – The most technologically, sophisticated commercial indoor farm in the world. The technology automatically generates ideal conditions for the plants, while collecting the data as they grow. The data help in providing the plants with the exact amount of light, nutrients and water.
12. Sky farm – A wind-powered vertical farming tower.
13. Sky Greens – The World's First Hydraulic Driven Vertical Farm. The Singapore-based company Sky Greens has developed a revolutionary vertical farming system which is also the world's first low carbon, hydraulic driven farm. The vegetables are planted on shelves that keep on rotating throughout the day.

Sustainable Vertical Farming Ventures in India

Mushroom Production

Mushrooms do not require lots of light to grow and are nutritious food that cycle up organic waste. Mushroom producers have been using vertical farming much longer than the plant producers. It is amenable to urban or indoor farming in vertical beds and is being practiced in several metros and peri-urban areas. Mushroom production is an apt example of successful, economical and sustainable vertical farming.

Poultry Production (Broiler and Eggs)

Poultry birds, mainly the layers (egg producers) and broilers (meat bird), are raised either in multi-storey buildings on floors or in battery cages. Battery cages accommodate more number of birds making the venture more profitable. Though popular both in urban and rural areas, the animal welfare organizations have serious reservations and do not support it.

Garden Wall/Bio Wall

Garden wall–a type of vertical farming (garden) is popular throughout the world including India. Vertical gardens can be seen on public places mainly on airports, metro pillars/stations, river bridges, elevated roads, etc., in many metro cities in India. Multicolour evergreen foliage plants are preferred, although flowers and even leafy vegetables find place in these wall gardens. However, their maintenance is costly and proving un-sustainable. Orchids, which do better on tree trunk and its branches, can be grown vertically under controlled climatic conditions especially in temperate and sub-temperate climate. Garden walls with seasonal leafy vegetables are the best option for indoor gardens. Availability of all season plants, growing media, fertilizers and watering are issues that need to be addressed for their commercialization.

Plants suitable for bio/ living walls:

For garden walls, selection of suitable plants is the single most important factor in its success. Plants thriving well in local climate are the best option. Besides, they should have compact growth providing thick and dense cover. Plants having short growth habit, shallow fibrous root system and longer life cycle are best suited for this purpose. They also need to be capable of coping both full sun and full shade according to the location. Most commonly used plants in bio/living walls are:

Green facades

Climbers having faster growth and dense variegated foliage from ground to top, are the best like *Hedera helix,Parthenocissu squinquefolia, Parthenocissustricuspidata (Japanese creeper), Hydrangea anomalapetiolaris (Climbing Hydrangea), Lonicera japonica, Jasminum officinale, Aristolochia spp., Clematis paniculata, Cissusrhom bifolia (Grape Ivy, Ficus pumila (Creeping fig), Bignonia unguiscatii (Cat's-Claw Vine) , Polygonumbauld schianicum, Passiflora caerulea,* etc.

Living wall

Plant species with dwarf nature, limited root volume, fibrous roots, resistant to wind, and good growth habit are ideal, e.g., *Dracaena marginata, Dracaena sanderiana, Phalaenopsis* spp., *Hibiscus* spp., *Gardenia* spp., *Nephrolepsis, Asparagus sprengeri, Kalanchoe* spp., *Fittonia* spp., *Scindapsus aureus, Asplenium nidus, Maranta* spp., *Cordyline* spp., *Chlorophytum* spp., *Haworthia* spp., *Tradescantia* spp., *Hydrangea, Nasturtium, Wisteria, Petunia, Clematis, Selaginella, Daisies, Bromeliads* and even few vegetables like tomato, chillies, cucumber, pea, lettuce, etc.

Exterior wall

Selected plants should perform well under full sun shine or partially shade conditions, e.g., *Lavendulaangus tifolia, Thymus serpyllum, Rosmarinus or Salvia splendens,* for full sunlight. *Rex begonia, Arum, Davallia, Asplenium nidus, Fuchia for* shady conditions.

Interior wall

Selected plants should perform well both under full shade or partially shade conditions, e.g., *Philodendron, Epipremnum, Aeschynanthus, Aeschynanthus, Columnea, Saintpaulia,* many species of *Peperomia and Begonia* or different ferns *like Nephrolepis* and *Pteris.*

Hydroponic Fodder

An age-old technology for the multiplication of green fodder to meet daily requirements of cattle is continuously being updated and popularized by public and private sectors. It involves climate-controlled cabins with trays vertically arranged and fed hydroponically through nutrient film technique. Barley, oat and maize are commonly used for hydroponic green fodder production. Not only the hydroponics technology for fodder production has been successfully developed, evaluated and certified by Ministry of Agriculture for various fodder crops, the fodder is being produced in Ayurvet Pro Green Hydroponics Machines. Fodder banks equipped with Certified Hydroponics Machine (both all-weather and make shift) need to be supported at village and farm levels for providing round the year (24 x 7) value added green fodder for animals.

Strawberry Vertical Farming

Strawberries are commercially grown in environment controlled structures in different gutter systems by keeping distance between two gutters approximately 100 cm. A gutter system enables correct amount of drainage water to irrigate.

Cucurbits Trained Vertically

Viny cucurbits spread on ground occupy huge space with low productivity. In vertical gardens, vines are trained to climb a trellis or garden netting, accommodating a larger plant population leading to multiple productivity under open field condition. Same is done under protected conditions in soil or soilless medium. It is best suited to indeterminate varieties of tomatoes, yard long cow pea and several beans. The important cucurbit crops for vertical training include cucumber, bottle gourd, sponge gourd, ridge gourd, snake gourd, and bitter gourd.

Micro Greens

Micro greens and baby greens are a relatively new specialty becoming popular among nutrients conscious urbanites. These crops consist of vegetables and herbs that are consumed at a young growth stage. Several species of micro greens also contain high concentrations of health-promoting phyto-chemicals. Micro greens can be produced by vertical farming indoor, outdoor, on and off season, in home and on farm. They are preferred by chefs and consumers to enhance the flavour, colour, and texture of a dish. Off late micro greens of the genus *Brassica* mainly broccoli have become popular due to their ease of germination, relatively short production time (7 to 21 days), and offering of intense flavours, colours and medicinal values. Commercial growers are currently producing micro greens in greenhouses in vertical system using soilless media in trays. Micro greens are also produced hydroponically using capillary mats placed in troughs, similar to those used in a nutrient film technique system.

Microgreens at DRDO-DIHAR, Leh

Aquaponics

Aquaponics combine the fishes and plants in one ecosystem, wherein fishes grow in indoor ponds producing nutrient-rich waste that acts as a food for the plants grown in vertical farms. The plants purify and filter the wastewater that is recycled directly to the fish ponds. There is need of using Re-circulatory Aquacultures System (RAS) in the Vertical Farming. The pisciponics - cultivating fishes, prawns and snails with plants (vegetables, fruits, herbs or flowers) in a symbiotic environment with RAS in vertical farming is coming up in Kerala. This closed-cycle system has potential to become popular in coming days.

Farm Containers

The soilless cultivation, mostly of leafy vegetables, in controlled environment can be done in mobile or static containers. The high cost and care are major impediments in popularization of such farming.

The Structures and other Engineering Inputs in Vertical Farming

Vertical farms need to be safe, efficient, durable, elegant, economical and ecologically sustainable. The structure should meet optimum physiological and behavioral needs of the intended life forms with adequate micro-climatic control and access to appropriate nutrition. Heating, ventilation, and air conditioning (HVAC), uniformity of the environment, optimal delivery of carbon dioxide (CO_2), shelf spacing and design and smart lighting are important engineering components in environment controlled vertical farms allowing optimum interaction of crops and surrounding climate. Since seed germination, rooting, vegetative growth, flowering, fruiting, flavour and colour development are directly related to photo period, its duration, intensity, spectrum, etc., are crucial to maximize yield and quality of produce. LED, red and blue light alone and their combinations have direct influence on the crop growth. The energy intensive lighting and HVAC systems account for nearly one third of the overalloperational costs. The growing beds are suitably designed to utilize maximum space combining ease of irrigation/fertigation and other operations.

Commonly Grown Crops in Vertical Farming

Although possible to grow any plant completely indoor, economics limits to few crops in vertical farming that have small growing habit (*for maximizing the number of plants that can be grown in a limited space*), are prolific producers (*e.g. indeterminate tomatoes*), or can be grown and sold rapidly (*such as micro greens*). Compatible crops for vertical agriculture today include lettuce, broccoli, spinach, chard, chive, palak (beet leaf), mustard greens, amaranths, parsley, coriander, mint, kale, basil and other herbs (rosemary, fennel, thyme, oregano and others), strawberries, mushrooms, micro greens and sprouts, summer squash, peppers, eggplants, tomatoes, cucumbers, muskmelon, algae, crop nurseries, ornamental foliage and flower plants.

Vegetable production DRDO-DIHAR, Leh

Nizammudeen Yamuna Bridge New Delhi

Growing Medium

The type of growing media, providing needed nutrition to the relatively shallow depth/limited volume in containers, is the most important component in vertical farming. Field soils are generally unsatisfactory for the production of plants in containers. Mixing coco peat, vermiculite and perlite in the ratio of 3:1:1 on volume basis was observed to be one of the best growing media. However, to bring down cost locally available inert media needs to be standardized.

Water and Nutrient Management Efficiency in Vertical Farming

In vertical farming, the elevation difference between first row (*bottom one*) and top row (*upper one*) is considerable and this difference affects the water emission rate from dripper. Since success depends on the relation between head and discharge, the network of pipes and dripper discharge need to be such that all plants get uniform water irrespective of their location. For quality production, irrigation water needs to be free from undesirable chemicals and pathogens, while fertigation can save as much as 60% fertilizers. Appropriate design of irrigation/ fertigation system is important to maintain the pressure and discharge requirements. Essentially the drainage and recirculation of water/ nutrients solution needs to be monitored for its electrical conductivity (EC) and pH.

Electrical Conductivity (EC) and pH

The EC and pH are two key important parameters for hydroponics' success. Although crop specific, the EC values range from 1.5 to 2.5 ds/m for hydroponics depending on environment. Higher EC increases osmotic pressure and hinders nutrient uptake while lower EC severely affects the plant growth and yield. Reduction in water uptake is strongly and linearly correlated to EC. The pH value determines the nutrient availability for the plants; therefore, it needs daily adjustment due to the lower buffering capacity of soilless systems. Regulation of pH can also be carried out by using nitric, sulphuric or phosphoric acid, either individually or in combination.

Aeration for the Roots – Not only oxygen is essential for cell growth and activity, the roots require oxygen to absorb water and nutrients. Therefore, aeration, is an important factor that influences root and plant growth in hydroponic system. Maintenance of proper balance of water and oxygen in plant roots is very important in vertical farming and requires standardization.

Vertical farming, farming up rather than out is becoming popular for which greenhouses space is better utilized. The infrastructure for vertical farming both for using soil and soilless medium is to be suitably designed for maximum benefit with ease.

Vertical farming being highly technical, requires intensive research on:

a. Designing cost effective structures for vertical farming.

b. Media for production of plants and suitable crop varieties.

c. Standardization of plant nutrients requirements for variety of crops.

d. Irrigation methodologies and water requirement of crops and their hybrids/ varieties in different methods of vertical farming.

e. Plant protection preventive and management practices measures.

f. Detailed study on the role of different lights indoor need to be undertaken for different crops and growing environments and season.

Thrust areas for future research on vertical farming are enumerated below:

- Conservation of more thermal radiation in winter.
- Durable infrastructure modelling.
- Development of successful commercial models.
- Modernization of poly houses to permit agro-climatic manipulations of higher magnitude.
- Identification of new crops/varieties suitable for cultivation inside the greenhouse.
- Socioeconomic impacts of greenhouse technology in hilly region/cold arid zone.
- Plastic films of suitable grades need to be made available in adequate quantity.
- Development of area-specific suitable greenhouse designs.
- Standardization of package of practices for growing various crops in greenhouse for specific regions.
- Varieties suitable for greenhouse cultivation need to be made available to enhance the productivity.
- Vocational training to the unemployed youth and greenhouse practitioners in remote areas to update their skills in this technology.
- Financial assistance such as low-interest loans should be made available for poor farmers to adopt this technology.
- Insurance of greenhouse structure and crop to protect the farmers against the loss caused by natural calamities.
- Rigorous extension program to disseminate the technology information to the poorest of the society.

The vertical farming has advanced abroad, providing positivity to space farming, which would be in demand even on other planets inhabited with either short or long periods.

B. Surface covered cultivation-use of plastic mulch

The development of polyethylene as a plastic film in 1938 and its subsequent introduction as a plastic mulch in the early 1950s revolutionized the commercial production of horticulture crops mainly the vegetable crops world over by following soil surface covered cultivation. Throughout the succeeding years, research, extension, and industry personnel, together with farmers, have documented the advantages of using plastic mulch in horticulture and other crops. Mulching provides a protective covering around plants to prevent evaporation of soil moisture, checking weed germination and growth which prevent stress on plant roots. This covering can also be of organic matter like straw, leaves, sawdust, peat and others as well as of inorganic matter like plastics. Here use of plastic mulch is discussed. When plastic is used as mulch material in crop management it is called plastic mulching or use of plastic as mulch. This chapter covers some experience of plastic mulch in horticulture crops in India. Plastic mulching is a process of covering the soil around plant root area with a view to insulate the plant and its roots from the effects of extreme environment fluctuations. Low density polyethylene (LDPE) and linear low density polyethylene (LLDPE) plastic films are used for mulching. Plastic mulch affects the micro-climate surrounding the plant through absorption, transmission and reflection of solar energy creating suitable temperature below and above the mulch. Plastic mulch should be air and thermal proof to check evaporation from soil and maintain the heat absorbed.

Major Role of Plastic Mulch

Based on experimentation and testing on farms the following role of plastic film in horticulture production is noticed.

*Moisture Conservation:*Plastic mulch suppresses water evaporation and enhances infiltration of water avoiding its scarcity and excess of.

*Soil Temperature:*During winter roots are protected from low temperature or cold. During summer heat transfer is prevented by using proper reflective colour mulch.

*Weed control:*Coloured plastic prevents light penetration to soil and checks weed seeds germination. If germinated the mulch layer is too thick or impenetrable for weed seedling to emerge.

*Control Soil Erosion:*Plastic mulch prevents free flow of water on soil surface avoiding soil erosion mainly on sandy and sandy loam soils. In case of rains water force is prevented by plastic film thus checks soil erosion.

*Disease control:*Mulch is a protective layer between soil and plant preventing soil borne disease transmission to plants, keeping produce off soil or suppressing disease growth in soil.

*Insect-pest control:*Mulch acts as barrier to insect keeping adults from entering the soil and laying eggs. Insects are kept at bay by reflection in case of particular color plastic mulch. But root- knot nematodes do multiply better under plastic mulch- a negative role of plastic mulch.

Benefits of Use of Plastic Mulch

Based on results of several investigations, the following benefits of use of plastic mulch for commercial crop production have been observed.

1. Moisture retention by checking evaporation from soil surface.
2. Inhibits weeds growth.
3. Facilitate drip irrigation.
4. Reduces fertigation leaching by rain fall or runoff during rainy season.
5. Decreases soil compaction which provides better environment for plant growth.
6. Protects fruit quality in vegetables by reducing rain-splashed soil deposits on fruit. In addition mulches also reduce fruit rot caused by soil borne pathogens.
7. Helpful in soil fumigation: mulches increase the effectiveness of soil fumigants.
8. Helps to control pests: Highly reflective mulches assist management of insect vectors of diseases.
9. Plastic mulches maintain a warm temperature during night which facilitates an early establishment of seedlings by strong root system or germination of seeds.
10. Soil water erosion is minimized.
11. Plastic mulches serve for longer period. They can be used for more than one season.
12. Plastic mulch helps in increasing root growth of the plant which contributes to early and higher yields in vegetable and flower crops.
13. It significantly improves plant stand establishment of many vegetable and flower crops.
14. It is cheaper and safer technique and involves no phytotoxicity or chemical residue.
15. Plastic mulch is farmer's and plant friendly.

Disadvantages of Using Plastic Mulch

1. Since plastic films are non- biodegradable it is not considered environment friendly farming practice.

2. Difficulties in the removal, upkeep, reuse and disposal of plastic mulch.
3. Specialized machinery is required for large scale use of plastic mulch and hence higher expenditure.
4. Clear plastic mulch promotes weed growth that in turn deprive both soil moisture and nutrients from the crop.

Useful Tips in Application of Plastic Mulch

In ornamental plant and fruit crop nurseries to check weed growth in working space it is advisable to apply plastic mulch mechanically. On small piece of land or kitchen gardens it can be applied manually. Apply FYM and fertilizers before mulching

The plastic mulch should be held tight touching the soil without any crease or fold on the bed. Borders (10cm) should be anchored inside the soil for about 7-10cm deep in small furrows at an angle of 45°.The mulch material should be punctured at the required distances as per crop spacing and laid on the bed. The seeds/seedlings should be sown/ transplanted in the holes/punctured place. Plastic mulch provides an insulation barrier between the soil and air, moderate the soil temperature.

Do not stretch the film very tightly. It should be loose enough to overcome the expansion and shrinkage conditions caused by temperature and the impacts of cultural operation. The slackness for black film should be more as the expansion; shrinkage phenomenon is maximum in this color. The film should not be laid on the hottest time of the day, when the film will be in expanded condition. High value crops can absorb expenditure on plastic mulch better. Water harvesting is better if channels and reservoir are lined with plastics. Vegetable crops like tomato, brinjal. Capsicum, chili, okra, cabbage, cauliflower, knol-khol, broccoli, squash, pumpkin, water melon, muskmelon, bottle gourd, ridge gourd, sponge gourd, chow-chow, parwal, snake gourd and others respond well to plastic mulch.

Selection of an Appropriate Plastic Mulch

Selection of plastic mulch depends upon the specific purpose to be achieved such as weed control, raising or lowering of soil temperature, disease and insect control, getting early or late crop. Different color plastic plays different role as has been explained in this article.

These days' plastic mulches are available in different colors and thickness keeping in view their multiple uses which are becoming popular among farmers. In plastic mulches the most preferred colours are transparent (clear) and black, although wide variety of shades and colours are being used for specific purpose in the production of food crops. Currently worldwide double color plastic mulches viz., sliver + black, blue + black, yellow + black, white + black and red + black are being tested in production of different horticultural crops and more specifically vegetables crops in different seasons.

Looking to above and other uses and advantages large scale use of different colour, plastic mulches is suggested in vegetable production. The use of 10-25micron thick plastic mulches is suggested for different vegetable crops in different seasons, areas keeping in view the objective

A mulch colour affects the temperature below and above the mulch through the absorption, temperature of air around the plant, transmission and reflection of solar energy, Soil salinity (due to lesser quantity of water used and due to reduction in evaporation and prevention of upward movement of water). This affects the micro environment surrounding the plants. The degree of contact between the mulch and the soil also affects the soil warming.

Different colour mulches having different role which are as follows:

1. *Black plastic*

A solid sheet of polyethylene effectively controls annual weeds. The disadvantage of black plastic is that water and oxygen cannot pass through this material. The soil should be moist prior to laying this synthetic mulch. It warms the soil and perennial weeds may be suppressed by black plastic. Black plastic films do not allow the sunlight to pass through on to the soil.

Photosynthesis does not take place in the absence of sunlight below black film. Hence, it arrest weed growth. Before lying of plastic mulch, the soil should be moist otherwise it warms the soil. All colour of plastic mulch control weed growth except clear plastic. Black, brown, clear plastic warms the soil, whereas white plastic cools the soil. Blue, green and silver plastic also produced greater yield than black plastic.

2. *Clear plastic*

This mulch which increases the soil temperature more than black plastic will not able to control weeds since sunlight can reach the soil surface. Soil temperature during the day time under clear plastic can reach 8-14° F higher at the 2" depth and 6-14° F higher at the 4" depth than bare soil at the same depths due to a greater (85-95%) solar radiation transmittance.

Clear plastic absorbs very little solar radiation. Water droplets that condense on the underside of clear plastic allow solar light (short wave radiation) in, but block outgoing, long wave infrared radiations (heat). This heat normally is lost to the atmosphere from bare soil.

Incoming solar radiation however makes weeds a major problem under clear plastic, unless controlled with a herbicide or fumigant.

3. *White plastic*

Light is reflected back into the atmosphere or the plant canopy from white plastic mulch, resulting in slightly cooler (-2° F at 1" depth) soil temperature. White plastic mulches can be used to establish crops in the summer when a reduced soil temperature might be beneficial.

4. *Silver/aluminium*

Reflective silver or aluminium mulches also give cooler soil temperature. They tend to repel aphids, which can serve as vector for various viral diseases.

5. *Red mulch*

According to the Virginia Corporation Extension, crops grown in red plastic produced up to 20% more tomatoes and boasted larger fruit size and weight over those grown in black plastic.

6. *Others*

Blue, yellow, orange, grey plastic mulches are also used. The different radiation patterns that are reflected back into the canopies of various crops from these mulches affect the plant growth and development in different ways. Some colours like yellow attract certain insects like green pea aphids and cucumber beetles. Therefore, such mulches might be used in a field to grow catch crops to pull insects away from other crops. Blue coloured mulches have been shown to increase honeydew yield. Black, silver + black, yellow + black, blue + black, white + black mulches greatly inhibit lights penetration to the soil, due to this weeds seeds do not germinate under these mulches. Clear or transparent mulch allows light penetration and weeds germination and growth. Insect vectots can repelled with colour of plastic. Basically yellow + black is used if the whitefly is the vector for viruses; silver + black is used in case aphid is the vector and blue + black is used if thrips are the vector for virus. Plastic film requirement can be calculating by referring following:

Table: Plastic mulch requirement

Thickness Micro Gauge Mm	Area Covered (Sqm/kg)	Weight (g/sqm)
7 28 0.007	144	6.9
20 80 0.02	54	18.4
25 100 0.025	42	23
40 160 0.04	26	38
50 200 0.05	21	46

Contribution of Plastic Mulch in Cold Desert of Ladakh

Till early 1990s cucurbits (gourds and melons) and solanaceous fruit vegetables (tomato, chili, capsicum and brinjal) cultivation in open field in Ladakh was considered impossible. The author experimented plastic mulch for raising above crops. By growing seedlings in poly pouch and plug trays under poly houses during April-May and transplanting them in open field (temperature 15-16 degree Celsius) with black coloured plastic mulch made it possible to grow them all in Leh, Kargil and Partapur valley. This has not only improved vegetable basket in the area but also added variety to food basket of local inhabitants and soldiers. Sharda melon imported in large quantity in the country can be produced in these areas with ease. Perhaps it is a major contribution of use of plastic mulch in high altitude to make possible cultivation of vegetables there for meeting food and nutrition requirements of locals locally.

Plastic mulch in different crops

Plastic mulch laying

New Age Production Technologies

Soilless Cultivation: Soilless agriculture is the cultivation of crops in media other than soil. This is new age crop production technology where plant growing media is other than soil such as rock wool, coco peat, perlite, vermiculite, sawdust, bark chips, sand, gravel, pumice, polyurethane mats etc. Button mushroom and orchids production under covers is an example of soilless cultivation. Water, plant nutrients and mixtures of the above are used as substrates for growing of plants indoor or outdoor. It is most suited for vertical gardens under protected conditions or balconies. This technique can make best use of waste land or sick soils roof tops and others. Water is a major and essential resource in soilless cultivation of vegetables. Difficulty and cost of controlling soil born pests and diseases, soil salinity, lack of fertile soil, water and land shortage etc., have led to the development of soilless substrates and technology. Soilless cultivation can be conducted within protected environment virtually anywhere because it does not rely on soil suitable for agriculture. It conserves land and water, not affected by environmental changes and requires lesser pesticides. There is an increased yield per unit of water, space and energy. It facilitates maximizing space utilization through vertical growing. Soilless cultivation is the future for an ever growing world population. One cannot afford to continuously plant and re-plant in the soil or risk their yields to soil deterioration and climate change. It has become a desirable practice in to have vegetable and other crop nurseries in plug trays multiplied in soilless medium under protected structures.

Late Dr APJ Abdul Kalam, Former President of India appreciating EPCOT "Experimental Prototype Community of Tomorrow" located in the Disney Land, Florida, USA; soilless vegetable production on 23 October 2009.

Vegetable Seedling Plugs: Most of the vegetables can be transplanted by raising their seedling plugs in soil less medium under small protected area well in advance (one month or so). The seedlings can be grafts also. Seedling plugsensure better stand in the field and help in regulating transplanting date reducing their time in main field/structures to avoid competition with weeds and protection from adverse weather.

Hydroponics : Cultivation of plants in nutrient solutions is known as hydroponic cutlivation. This ia gerally done under protected structures and in containers arranged horizontally or vrtically (vertical garden). Use of nutrient film or fertilizers is made to feed the plants. Hydroponics can be practiced following nutrient film technique method, trough method, deep water floating method and coco peat bag culture.

Advantages of hydroponics

One can grow plants anywhere. It uses 20 times less water than soil based cultivation and 20 per cent less space for growing. It requires less of water and pesticides. Water and waste nutrients can be reused. No intercultural operation like weeding, tilling, mulching, change of soil etc. are required

Disadvantages of hydroponics

Putting together a hydroponic system is expensive. Constant supervision is required. Hydroponic systems are vulnerable to power outages. Growing a hydroponic garden demands technical expertise. Large scale production yet to be standardized.

Aeroponics: This is a form of hydroponics. Nutrient and water mixture is sprayed in caged roots under dark at certain intervals with the help of sensors. This conserves water, space and is less cumbesome. It demands constant supervision and has disadvantages as listed under hydroponics.

Aquaponics : Involves production of fish and vegetables with same input resource of water and nutrients. Feed is given to the fishes in the pond not to the vegetables or plants grown as hydroponics. Fish pond or tank water is used as hydroponics or areoponics with or without additional fertigation. This results in effective utilization of fish pond water and its nutrients for production of vegatables/plants simultaneously with added returns on input resources. Plants act as biofilter for the recycled water to fish pond.

Vegetable grafts: Vegetable grafts with resistant root stock and high as well as quality yield scion are becoming popular to manage biotic and abiotic stress. Even though grafting originated in East Asia (Japan, 1920), could not get attention of South Asia but became popular in European countries where greenhouse cultivation is not uncommon. Grafted plants are produced to a tune of several million tons in Spain, France and Italy. In Japan 93 % watermelon and 73% cucumber area of total area covered are grafts. AVRDC- Asian Vegetable Research and Development Centre, Taiwan recommends growing tomato on eggplant accessions EG 195 and EG 203 and sweet pepper on chili accessions PPO273-7502 and PPO 242-62 and Lee B for flood tolerance. In China 20% area under watermelon and cucumber is covered with their grafts. In India sporadic attempts have been made and resistant/tolerant rootstocks have been identified. Most common bio-stress in vegetables such as root-knot nematodes (solanaceae and cucurbitaceae family vegetables), *Fusarium* wilt (melons and tomato), bacterial wilt (tomato and eggplant), *Verticilium* wilt (tomato), *Phytophthora* (sweet pepper) are being solved by using resistant root-stocks. In vegetable crops wilt and root-knot nematode resistant in Solanaceous vegetables;*Phytophthora* resistant for chilli and sweet pepper;*Fusarium* wilt resistant for cucurbits;salt tolerant for cucurbits; cold tolerant for cucumber, muskmelon and bitter gourd rootstocks have been developed in India. Rootstock breeding is now picking up with scion compatibility traits in vegetable and other crops.

List of Greenhouses Manufacturers

Agriplast Tech India Pvt Ltd; Survey no. 426/3B-1B, Nallure Village; opp Nellure Govt. High School, Hosur Panchayat Union and Taluk , Krishnagiri Dist-635103; Tamil Nadu; www.agriplast.co.in 08141446666.

Classic Agricon, SCO No. 394, Ist Floor, Cabin No. 12, Sector -20, Panchkula; Haryana-134112; 09876427998

Essen, Mutipack Ltd, Survey No. 209. Plot no. 5. Industrial Area, Veraval (Shapur), Rajkot-360002, Gujarat

Future Farms; 4/640, 12 Link III Cross Street, Ist Main Road, Nehru Nagar, Kottivakam, Chennai-600041; www.futurefarms.in ; 08137848383/09743219388

Green pot Organic; House No 216, Sector-29; Faridabad; Haryana-121008; Greenpotorganic33 @gmail.com

Greenhouse Manufactures Association (IGMA). Plot No: 49, MIDC Floriculture park, Talegaon Dhabade, Tal Maval, Dist Pune – 401507 info@igma.org.in | www.igma.org.in Registration No: 29138

Indian Greenhouses Pvt Ltd; 7, Namrata Commercial Centre, Talegaon, Dabhade(station)Tal. Maval Dist., Pune-410507, Maharashtra;

Jain Irrigation Systems Ltd;Jain Plastic Park, NH No.6 Bambhori, Jalgaon 425001.Maharashtra, India..

Netafim Irrigation India (P). Ltd Ist Floor 297 CST Road, Vijaynagari, Kalina, Mumbai-400098; 022-2695 2761.

Netafim Irrigation Pvt Ltd; S1 and S10, Plot No. 16; Pankaj Arcade, 2nd Floor; Dwarka Sector, 5; New Delhi-110075

Pet Bharo; 98 , Bendre Nagar, Dharwar,Karnataka- 580008; www.petbharoproject.co.in

Rajdeep Agri Products (P) Ltd; 3229/1, Ranjeet Nagar; New Delhi: 110008; 011-2574881

Saveer Biotech Limited; 1442, Wazir Nagar, New Delhi-110003; bisiness@saveer.com; 01124620211

Bibliography

Abdul Hakkim V.M, Abhilash Joseph, E., Ajay Gokul AJ, Mufeedha K. 2016. Precision Farming: The Future of Indian Agriculture. J App Biol Biotech. 2016; 4 (06): 068-072. DOI: 10.7324/JABB.2016.40609

Adams, P. 1994. Nutrition of greenhouse vegetables in NFT and hydroponic system. *Indian Horticulture*. 361:245-257

Alderman, S. 2015. "The Practicality and Sustainability of Aquaponic Agriculture versus Traditional Agriculture with Emphasis on Application in the Middle East." Undergraduate thesis. Texas State University, San Marcos, US.

Al-Kodmany, K. 2018. The vertical farm: A review of developments and implications for the vertical city. *Buildings* 8 (2): 24.

Ambreen Nabi, Dr Sumati Narayan, Baseerat Afroza, Faheema Mushtaq,Shahnaz Mufti, Ummyiah HM and Ajaz Malik 2017. Precision farming in vegetables. *Journal of Pharmacognosy and Phytochemistry*: 6(6): 370-375

Ancha Srinivasan. 2006.Handbook of Precision Agriculture: Principles and Applications; CRC Press pp. 683 ISBN 9781560229551

Angmo, P., Dolma, T.,Namgail, D., Tamchos, T., Norbu, T.,Chaurasia, O. P. and Stoban, T. 2019 Passive Solar Greenhouse For Round The Year Vegetable Cultivation I Trans- Himalayan Ladakh Region, India. *Defence Life Science Journal*. 4(2): 103-116.

Asia Hiroyuki Takeshima and Pramod Kumar Joshi 2019. IFPRI Discussion Paper 01814 March 2019. Protected Agriculture, Precision Agriculture, and Vertical Farming Brief Reviews of Issues in the Literature Focusing on the Developing Region in Asia.

Bala, B. 2013. Upliftment of rural economy through protected cultivation. *Global Research Analysis*. 2 (2):42-43.

Buckseth Tanuja, Sharma, A.K., Pandey, K.K., Singh, B.P. and Muthuraj, R., 2016. Methods of pre-basic seed potato production with special reference to aeroponics- A review. *Sci Horti*., 204:79–87.

Chandra, P. and Panwar, J. 1987. Greenhouse technology and its scope in India. In: Proceedings of the national symposium on use of plastics in agriculture, New Delhi, pp 62–66

Chandra, P. 1976. Predicting the effects of greenhouse orientation and insulation on energy conservation. Unpublished MS thesis. Department of Agricultural Engineering, University of Manitoba, Winnipeg, Manitoba Canada. 95 p.

Chandra, P. 1994. Status of greenhouse development in India, Proceedings of the XXIX Annual Convention of Indian Society of Agricultural Engineers. Junagadh Agricultural University, Junagadh, Gujarat.

Chandra, P. Sirohi, P. S., Behera, T. K. and Singh, A.K. 2000. Cultivating vegetables in Polyhouse. Indian Horticulture. ICAR Publication.

Das, P.K. and Gope, H.L. 2014. Wireless sensor network technology for agricultural development: Challenges and chances in Bangladesh. *International Journal of Scientific & Engineering Research* 5(8): 162-167.

Despommier, D. and Carter, M. 2011. The Vertical Farm: Feeding the World in the 21st Century. UK: Picador.

Development of protected cultivation in Haryana. Working Group Report, Haryana Kisan Ayog, Panchkula, Haryana. 2014. 68p

Dixit, J., Dixit, A.K., Lohan, S.K. and Kumar, D. 2014. "Importance, Concept and Approaches for Precision Farming in India." In *Precision Farming: A New Approach*, edited by P. Singh, R. Singh, S.K. Lohan, T. Ram, 12-35. New Delhi: Daya Publishing House, New Delhi.

Ghosh, Arupratan. 2019. Greenhouse technology, NIPA, Pitam Pura, New Delhi, 326 pp

Gondkar, S., Singh, B.K. and Rao, D.U.M. 2016. Assessment of extent of entrepreneurial success among the protected agriculture entrepreneurs. *Journal of Community Mobilization and Sustainable Development* 10(1): 53-57.

Greenhouse Solar Dryers https://www.researchgate.net/publication/286417592_Greenhouse_Solar_Dryers

Gundu R. 2017. Design and development of low cost greenhouse to raise different cultivars. *International Journal of Agricultural Science and Research*. 7(3):29-36.

Hanan, J.J., Honey, W.D., Goldsberry, K.L. 1978 Greenhose management.Springer, New York

http://agritech.tnau.ac.in/horticulture/horti_Greenhouse%20cultivation.html

Janke, R.R., Altamimi, M.E. and Khan, M. 2017. The use of high tunnels to produce fruit and vegetable crops in North America. *Agricultural Sciences* 8(07): 692-715.

Kalantari, F., Tahir, O.M., Lahijani, A.M. and Kalantari, S. 2017. A review of vertical farming technology: A guide for implementation of building integrated agriculture in cities. *Advanced Engineering Forum,* 24: 76-91.

Kalubhai, P. 2015. "Present Status and Prospects of Micro Irrigation System in Mahisagar District of Gujarat." Doctoral dissertation, Anand Agricultural University, Anand, India.

Kushwaha, M., and V.R. Raghuveer. 2017. Survey of impact of technology on effective implementation of precision farming in India. *International Journal on Recent and Innovation Trends in Computing and Communication* 5(6):1300-1310.

Maheswari, R. Ashok, K.R. and Prahadeeswaran, M. 2008. Precision Farming Technology, Adoption, Decisions and Productivity of Vegetables in Resource-Poor Environments. Agricultural Economics Research Review 21 (Conference Number): 415-424

Maheswari, R., Ashok, K.R. and Prahadeeswaran. M. 2008. Precision farming technology, adoption decisions and productivity of vegetables in resource-poor environments. *Agricultural Economics Research Review* 21: 415-424.

Mahmud, M.S.A., Buyamin, S., Mokji, M.M. and Abidin, M.Z. 2018. Internet of things based smart environmental monitoring for mushroom cultivation. *Indonesian Journal of Electrical Engineering and Computer Science* 10 (3): 847-852.

Mandal, S.K. and Maity, A. 2013. Precision farming for small agricultural farm: Indian scenario. *American Journal of Experimental Agriculture* 3 (1): 200.

Mondal, P. and Basu, M. 2009. Adoption of precision agriculture technologies in India and in some developing countries: Scope, present status and strategies. *Progress in Natural Science* 19 (6): 659-666.

Mondal, P., Basu, M., Bhadoria, P.B.S., Emam, A.A., Salih, M.H. and Adegbite, A.A. 2011. Critical review of precision agriculture technologies and its scope of adoption in India. *American Journal of Experimental Agriculture* 1 (3): 49-68.

Moriyama H., Sase, S., Okushima, L. and Ishii M. 2015. Which design constraints apply to a pipe framed greenhouse? from perspective of structural engineering, meteorological conditions, and wind engineering. *JARQ* 49(1):1-10

Moustafa, A.T. 2010. Potential of protected agriculture and hydroponics for improving the productivity and quality of high value cash crops in Qatar. ICARDA.

NAAS, 2019. Vertical Farming. Policy Paper. 89. National Academy of agricultural Sciences, New Delhi. 20pp

NAEDB 1992 Action plans on cold deserts: an integrated approach for sustainable development.NAEDB, UHF, Solan – 173230 (HP)

Natikar, P.K., Balikai, R.A. and Anusha, CH. 2016. Pest management strategies in precision farming. *J. Exp. Zool. India* 19 (1)1-8

Naved Sabir and Balraj Singh, 2013 Protected cultivation of vegetables in global arena: A review. *Indian Journal of Agricultural Sciences* 83 (2): 123–35

NCPAH. 2011. Greenhouse: A Reference Manual. National Committee on Plasticulture Applications in Horticulture. Department of Agriculture and Cooperation, Ministry of Agriculture, Government of India. 10th floor, International Trade Tower, Nehru Place, New Delhi 110019. NCPAH/TB/2010-

Ojha, T., Misra, S. and N.S. Raghuwanshi. 2015. Wireless sensor networks for agriculture: The state-of-the-art in practice and future challenges. *Computers and Electronics in Agriculture* 118: 66-84.

Pant, T., Agarwal, A., Bhoj, A.S., Joshi, R.P., Om Prakash and Dwivedi, S.K. 2018. Vegetables cultivation under hydroponics in Himalayas- Challenges and opportunities. *Defence Science J.*, 3(2):111-115.

Parvatha Reddy, P. 2016. Sustainable crop protection under protected cultivation. Spinger E-book. 433p.

Pascual, M.P., Lorenzo, G.A. and Gabriel, A.G 2018. Vertical farming using hydroponic system: Toward a sustainable onion production in Nueva Ecija, Philippines. *Open Journal of Ecology.* 8 (01): 25.

Patil, V.C., Nadagouda, B.T. and Al-Gaadi, K.A. 2013. Spatial variability and precision nutrient management in sugarcane. *Journal of the Indian Society of Remote Sensing.* 41 (1):183-189.

Paustian, M., and Theuvsen, L. 2017. Adoption of precision agriculture technologies by German crop farmers. *Precision Agriculture* 18 (5): 701–716.

PDF Precision Irrigation: Sensor Network Based Irrigation. Available from: https://www.researchgate.net/publication/221927923_ Precision_Irrigation_ Sensor_Network _Based_Irrigation [accessed May 04 2019].

Pinstrup-Andersen, P. 2018. Is it time to take vertical indoor farming seriously? *Global Food Security* 17: 233-235.

Prathyusha, K., Bala, G.S. and Ravi, K.S. 2013. A real-time irrigation control system for precision agriculture using WSN in Indian agricultural sectors. *International Journal of Computer Science, Engineering and Applications* 3 (4): 75.

Precision Irrigation: Sensor Network Based Irrigation. Available from: https://www.researchgate.net/publication/221927923_Precision_ Irrigation_Sensor_ Network _Based_Irrigation

Putri, N.Y., Sharfina, N.P. and Prakarti, T. 2015. Sky farming: The alternative concept of green building using vertical landscape model in urban area as an effort to achieve sustainable development. *International Journal of Civil, Environmental, Structural, Construction and Architectural Engineering* 9 (7): 938-941.

Ramamoorthy, R.R. and Gopu, A. 2016. "An Over View of the Implementation of Precision Farming Projects in Tamil Nadu, India." MPRA Paper No. 73674. Accessed February 7, 2019. https://mpra.ub.uni-muenchen.de/73674/1/MPRA_paper_73674.pdf.

Rango, A., Laliberte, A.S., Herrick, J.E., Winters, C., Havstad, K., Steele, C. and Browning, D. 2009. Unmanned aerial vehicle-based remote sensing for rangeland assessment, monitoring, and management. *Journal of Applied Remote Sensing* 3 (033542): 1-15.

Sangpradit, K. 2014. Study of the solar transmissivity of plastic cladding materials and influence of dust and dirt on greenhouse cultivations. *Energy Procedia* 56:566-573.

Shruthi, K., Hiremath, G. M., Joshi, A.T. and Patil, S.S. 2018. An economic analysis of precision agriculture—A case study of paddy in North Eastern Karnataka. *Indian Journal of Economics and Development* 14 (2): 274-280.

Shruthi, K., Hiremath, G. M. and Joshi, A.T. 2018. An application of precision farming techniques in production of pigeonpea—An economic analysis. *Progressive Agriculture* 18 (1): 15-19.

Singh, B., Dwivedi, S.K., Paljor, Eli 1998a. Studies on suitability of various structures for winter vegetable production at subzero temperatures (Abstr). In: 25th Indian horticulture congress, Belgium, p 290.

Singh, B., Dwivedi, S.K., Sharma, J.P. 2000. Greenhouse technology and winter vegetable production in cold arid zone. In: Sharma JP, Mir AA (eds) Dynamics of cold and agriculture. Kalyani Publishers, New Delhi, pp 279–293

Singh, B., Dwivedi, S.K., Singh, N., Paljor, Eli 1998b. Sustainable horticultural practices for cold arid areas. In: International symposium on sustainable agriculture in Hill areas (Abstr). Himachal Pradesh Krishi Vidyapeeth, Palampur, pp 35–36

Singh, B. 2016. New Age Crop Production Technology to Boost Farmer's Income. Indian Horticulture: 61(6), ICAR, Publication: 65-71.

Singh, B. 2016. Protected Cultivation with Special Reference to Vegetable Production. Proceedings XII Agricultural Science Congress, 03-06 Feb. 2015. National Dairy Research Institute, Karnal, Haryana, India. 383-394

Singh, B. 2017. Elements and Strategic Measures in Maximizing Production through Protected Cultivation of Vegetables towards Sustainable Horticulture Development in Asian Countries; "New Pathways to Sustainable Horticulture Development (NPSHD)". PNASF Publication.

Singh, B. and Dhaulakhandi, A. B. 1998. Application of solar greenhouse for vegetable production in cold desert. In: Proceedings of world renewable energy congress. Florence. Italy. 2311-14- V

Singh, B. 2016. Innovative protected horticulture and its potential for doubling farmer's income through horticulture. Horticulture Society of India Publication, pp 543-556

Singh, B., Singh, Balraj, Sabir, N. and Hasan Murtaza. 2014. Advances in protected cultivation. NIPA, New Delhi 248 p.

Singh, Brahma 2016. New Age Protected Cultivation, e- magazine, ISPC, New Delhi publication, Vol 2(1):41 p

Sirohi, P.S. and Behera, T.K. 2000. Protected cultivation and seed production in vegetables, *Indian Horticulture*, p23.

Soman, S., Byju, G. and R. Bharathan. 2013. GIS based decision support system for precision farming of cassava in India. *Acta Biologica Indica* 2 (2): 394-399.

Somerville, C., M. Cohen, E. Pantanella, A. Stankus, and A. Lovatelli. 2014. *Small-Scale Aquaponic Food Production: Integrated Fish and Plant Farming*. FAO Fisheries and Aquaculture Technical Paper No. 589. Rome: Food and Agriculture Organization of the United Nations.

Sopan, G.S. 2011. "A Critical Analysis of Entrepreneurs in Protected Agriculture in Maharashtra." Doctoral dissertation, Division of Agricultural Extension, Indian Agricultural Research Institute, New Delhi.

Spehia, R.S. 2015. Effect of poly house on productivity of vegetable crops as compared to open cultivation. *Current Science*. 108 (2254): 2254-7

Sreedhara D.S., Kerutagi M.G., Basavaraja, I., Kunnaland, H.L.B. and Dodamani, T. 2013. Economics of capsicum production under protected conditions in Northern Karnataka. *Karnataka J. Agric. Sci.,* 26(2):217-219

Srivastava, R., Sarkar, D., Mukhopadhayay, S.S., Sood, A., Singh, M., Nasre, R.A. and Dhale, S.A. 2015. Development of hyperspectral model for rapid monitoring of soil organic carbon under precision farming in the Indo-Gangetic Plains of Punjab, India. *Journal of the Indian Society of Remote Sensing* 43(4):751-759.

Subramanian, K.S. and Tarafdar, J.C. 2011. Prospects of nanotechnology in Indian farming. *Indian Journal of Agricultural Sciences* 81(10): 887-93.

Sun, H. and Li., M. 2015. "Precision agriculture in China: Sensing technology and application." In *Precision Agriculture Technology for Crop Farming*, edited by Q. Zhang. Boca Raton, FL, US: CRC Press.

Suparwoko and B. Taufani. 2017. Urban farming construction model on the vertical building envelope to support the green buildings development in Sleman, Indonesia. *Procedia Engineering* 171: 258-264.

Tiwari, G. N. 2003. Plastic greenhouses for controlled environment. Narosa Publishing House, New Delhi.

Tiwari, G. N. and Goyal, R. K. 1998. Greenhouse Technology. Narosa Publishing House, New Delhi. 252-311.

Vinod, P.G. 2017. Development of topographic position index based on Jenness algorithm for precision agriculture at Kerala, India. *Spatial Information Research* 25 (3):381-388.

Working group report on development of protected cultivation in Haryana. Haryana Kisan Ayog (www,haryanakisanayog.org)

Yadav, R.K., Kalia, P., Choudhary, H., Zakir Husain and Brihama Dev 2014. Low-Cost Poly house Technologies for Higher Income and Nutritional Security. *International Journal of Agriculture and Food Science Technology*: 5 (3): 191-196

Yadav, V., and P. Yadav. 2014. Precision farming: A sustainable approach for organic horticulture production. *Indian Horticulture Journal* 4 (1): 79-72.

Yaghoubi, S., Akbarzadeh, N.A., Bazargani, S.S., Bazargani, S.S., Bamizan, M. and M.I. Asl. 2013. Autonomous robots for agricultural tasks and farm assignment and future trends in agro robots. *International Journal of Mechanical and Mechatronics Engineering.* 13(3):1-6.

Zhang, C. and Kovacs, J.M. 2012. The application of small unmanned aerial systems for precision agriculture: A review. *Precision Agriculture* 13(6): 693-712. Zhou, J., X. Wang, R.

Zhang, Q. Feng, and W. Ma. 2012. "Automatic Navigation Based on Navigation Map of Agricultural Machine." In: *Computer and Computing Technologies in Agriculture VI*, edited by D. Li and Y. Chen, 304-311. Heidelberg, Germany: Springer.

FSC
www.fsc.org
MIX
Papier | Fördert
gute Waldnutzung
FSC® C083411

Kolibri 360 GmbH
Ferdinand-Jühlke-Straße 7
99095 Erfurt, Deutschland
produktsicherheit@kolibri360.de